# COSMIC EVOLUTION

# COSMIC EVOLUTION

## THE RISE OF COMPLEXITY
## IN NATURE

## ERIC J. CHAISSON

HARVARD UNIVERSITY PRESS
Cambridge, Massachusetts
London, England
2001

Illustrated by Lola Judith Chaisson
Designed by Gwen Nefsky Frankfeldt

*Library of Congress Cataloging-in-Publication Data*

Chaisson, Eric.
Cosmic evolution : the rise of complexity in nature /
Eric J. Chaisson ; illustrated by Lola Judith Chaisson.
p. cm.
Includes bibliographical references and index.
ISBN 0-674-00342-X (cloth : alk. paper)
1. Cosmology.   2. Life—Origin.   3. Matter—Constitution.
I. Chaisson, Lola Judith, ill.   II. Title.

QB981 .C412 2001
523.1—dc21      00-058043

In memory of H. Dudley Wright,
inventor, entrepreneur, interdisciplinarian
*par excellence*

# CONTENTS

# PREFACE

Using astronomical telescopes and biological microscopes, among a virtual arsenal of other tools of high technology, modern scientists are weaving a thread of understanding spanning the origin, existence, and destiny of all things. Now emerging is a unified scenario of the cosmos, including ourselves as sentient beings, based on the time-honored concept of change. From galaxies to snowflakes, from stars and planets to life itself, we are beginning to identify an underlying, ubiquitous pattern penetrating the fabric of all the natural sciences—a sweepingly encompassing view of the order and structure of every known class of object in our richly endowed Universe. We call this subject "cosmic evolution."[1]

Recent advances throughout the sciences suggest that all organized systems share generic phenomena characterizing their emergence, development, and evolution. Whether they are physical, biological, or cultural systems, certain similarities and homologies pervade evolving entities throughout an amazingly diverse Universe. How strong are the apparent continuities among Nature's historical epochs and how realistic is the quest for unification? To what extent might we broaden conventional evolutionary thinking, into both the pre-biological and post-biological domains? Is such an extension valid, merely metaphorical, or just plain confusing?

For many years, during the 1970s and 80s at Harvard University, I taught, initially with George B. Field, an introductory course on cosmic evolution that sought to identify common denominators bridging a wide variety of specialized science subjects—physics, astronomy, geology, chemistry, biology, and anthropology, among others. The principal aim of this interdisciplinary course explored a universal framework against which to address some of the most basic issues ever contemplated: the origin of matter and the origin of life, as well as how radiation, matter, and life interact and change with time. Our intention was to help sketch a grand evolutionary synthesis that would better enable us to understand who we are, whence we came, and how we fit into the overall scheme of things. In doing so, my students and I gained a broader, integrated knowledge of stars and galaxies, plants and animals, air, land, and sea. Of paramount import, we learned how the evident order and increasing complexity of the many varied, localized structures within the Universe in no way violate the principles of modern physics, which, prima facie, maintain that the Universe itself, globally and necessarily, becomes irreversibly and increasingly disordered.

Beginning in the late 1980s while on sabbatical leave at the Massachusetts Institute of Technology, and continuing for several years thereafter while on the faculty of the Space Telescope Science Institute at Johns Hopkins University, I occasionally offered an advanced version of the introductory course. This senior seminar attempted to raise substantially the quantitative aspects of the earlier course, to develop even deeper insights into the nature and role of change in Nature, and thus to elevate the subject of cosmic evolution to a level that scientists and lay persons alike might better appreciate. This brief and broadly brushed monograph—written mostly in the late 1990s during a stint as Phi Beta Kappa National Lecturer, and polished while resuming the teaching at Harvard of my original course on cosmic evolution—is an intentionally lean synopsis of the salient features of that more advanced effort.

Some will see this work as reductionistic, with its analytical approach to the understanding of all material things. Others will regard it as holistic, with its overarching theme of the whole exceeding the sum of Nature's many fragmented parts. In the spirit of complementarity, I offer this work as an evolutionary synthesis of both these methodologies, integrating the deconstructionism of the former and the constructivist tendencies of the latter. Openly admitted, my inspiration for writing this book has been Erwin Schrödinger's seminal little tract of a half-century ago, *What is Life?*, yet herein to straighten and extend the analysis to include all known manifestations of order and complexity in the Universe. No attempt is made to be comprehensive insofar as details are concerned; much meat has been left off the bones. Nor is this work meant to be technically rigorous; that will be addressed in a forthcoming opus. Rather, the intent here is to articulate a skeletal précis—a lengthy essay, really—of a truly voluminous subject in a distilled and readable manner. To bend a hackneyed cliché, although the individual trees are most assuredly an integral part of the forest, in this particular work the forest is of greater import. My aim is to avoid diverting the reader from the main lines of argument, to stay focused on target regarding the grand sweep of change from big bang to humankind.

Of special note, this is not a New Age book with mystical overtones however embraced or vulgarized by past scholars, nor one about the history and philosophy of antiquated views of Nature. It grants no speculation on the pseudo-science fringe about morphic fields or quantum vitalism or interfering deities all mysteriously affecting the ways and means of evolution; nor do we entertain epistemological discussions about the limits of human knowledge or post-modernist opinions about the sociological implications of science writ large. This is a book about mainstream science, pure and simple, outlining the essence of an ongoing research program admittedly multidisciplinary in character and colored by the modern scientific method's unavoidable mix of short-term subjectivity and long-term objectivity.

In writing this book, I have assumed an undergraduate knowledge of natural science, especially statistical and deterministic physics, since as we shall see, much as for classical biological evolution, both chance and necessity have roles to play in all evolving systems. The mathematical level includes that of integral calculus and differential equations, with a smattering of symbolism throughout; the units are those of the centimeter-gram-second (cgs) system, those most widely used by practitioners in the field. And although a degree of pedagogy has been included when these prerequisites are exceeded, some scientific language has been assumed. "The book of Nature is written in the language of mathematics," said one of my two intellectual heroes, Galileo Galilei, and so are parts of this one. Readers with unalterable math phobia will benefit from the unorthodox design of this work, wherein the "bookends" of Prologue-Introduction and Discussion-Epilogue, comprising more than half of the book, can be mastered without encountering much mathematics at all.

What is presented here, then, is merely a sketch of a developing research agenda, itself evolving, ordering, and complexifying—an abstract of scholarship-in-progress incorporating much data and many ideas from the entire spectrum of natural science, yet which attempts to surpass scientific popularizations (including some of my own) that avoid technical lingo, most numbers, and all mathematics. As such, this book should be of interest to most thinking people—active researchers receptive to an uncommonly broad view of science, sagacious students of many disciplines within and beyond science, the erudite public in search of themselves and a credible worldview—in short, anyone having a panoramic, persistent curiosity about the nature of the Universe and of our existence in it.

I thank those who read the penultimate draft of the manuscript, thereby saving me some embarrassing errors: Kate Brick, Larry Edwards, George Field, Dudley Herschbach, Jonathan Kenny, Hubert Reeves, Fred Spier, George Whitesides, and Rich Wolfson. Each of these distinguished specialists necessarily examined such an interdisciplinary work from a different perspective, and none of them can be expected to agree with all that remains—which is exactly the way that modern science seeks to discriminate sense from nonsense, selecting and accumulating the former at the expense of the latter and thereby moving us all toward a better approximation of reality.

Eric J. Chaisson
Concord, Massachusetts

# OVERVIEW OF COSMIC EVOLUTION

[We are] made of the same stuff of which events are made. . . The mind that is parallel
with the laws of Nature will be in the current of events, and strong with their strength.

—Ralph Waldo Emerson, from the essay "Power" in *The Conduct of Life*

Since the dawn of civilization, men and women have wondered
about, and even feared, the mysteries of the skies. At first, they approached their world subjectively, believing Earth to be the stable hub
of the Universe, with Sun, Moon, and stars revolving about it. Stability
led to a feeling of security, or at least contentment—a belief that the origin and destiny of the cosmos were governed by the supernatural.

With the advent of recorded history, our ancestors became aware of
another mystery, namely, themselves. Indeed, the origin and destiny of
human beings were as enigmatic as anything in the depths of space. Religions and philosophies held forth, providing grand myths, epic stories, and a genuine sense of well-being before an uncertain future.

Later, but only within the past few hundred years, humans began to
adopt a more critical stance toward themselves and the Universe, seeking to view our world more objectively. With it, modern science was
born, the first major product of which was the Copernican revolution.
The idea of the centrality of Earth was demolished forever, and with it
the false serenity that had been engendered by the unknown. Humankind came to feel that it was marooned on a tiny particle of dust drifting aimlessly through a hostile Universe.

More recent scientific developments, particularly within the past few
decades, have continued to suggest that, as living creatures, we inhabit
no unique place in the Universe at all. We live on what appears to be an
ordinary rock called Earth, one planet orbiting an average star called
the Sun, one stellar system near the edge of a huge collection of stars
called the Milky Way, one galaxy among countless billions of others
spread throughout the observable abyss called the Universe.

It is perhaps a sobering thought that we seem so inconsequential in the Universe. It is even more humbling at first—but then wonderfully enlightening—to recognize that evolutionary changes, operating over almost incomprehensible space and nearly inconceivable time, have given birth to everything seen around us. Scientists are now beginning to decipher how all known objects—from atoms to galaxies, from cells to brains, from people to society—are interrelated. We are attempting to sketch the unifying scenario of cosmic evolution, a powerful new epic for the new millennium.

Simply defined, cosmic evolution is the study of change—the vast number of developmental and generative changes that have accumulated during all time and across all space, from big bang to humankind. To quote some long-forgotten wit, "Hydrogen is a light, odorless gas which, given enough time, changes into people." More seriously and specifically, cosmic evolution comprises the sum total of all the many varied changes in the assembly and composition of radiation, matter, and life throughout the history of the Universe. These are the changes that have produced our Galaxy, our Sun, our Earth, and ourselves.

Tritely stated, though no less true, the word "evolution" implies neither dogmatism nor atheism. It harbors no premeditated, a priori implication for any religion or preferred philosophy of antiquity; there is no hidden agenda here. As used in this book, evolution is hardly more than a fancy word for change, especially, again, both developmental and generative change. The term itself derives from the Latin *evolvere*, meaning to unfold, to roll out. Indeed, it seems that change is the hallmark for the origin, maintenance, and fate of all things, animate or inanimate.

*Change:* To make different the form, nature, and content of something. Change has, over the course of time and throughout all space, brought forth, successively and successfully, galaxies, stars, planets, and life. Evidence for that change is literally everywhere. Whether we look out into the macroscopic realm with astronomical telescopes or down into the mesoscopic domain with biological microscopes or even sub-microscopically with high-energy accelerators, the most common feature perceived is change. Much of that change is subtle, such as when the Sun fuses sedately at mid-career for billions of years or when Earth's tectonic plates drift sluggishly across the face of our planet for equally long durations. By contrast, some of that change is much more dramatic, such as when very massive stars (unlike our Sun) perish catastrophically in supernova explosions or when geologic pressures amass near Earth's surface to cause sudden quakes and volcanoes.

Nothing seems immutable, nothing at all, much akin to the ancient philosophers' notion of "becoming" as a more genuine representation of existence. Indeed, Heraclitus of old may well have been right when he so cogently claimed some 25 centuries ago that there is nothing permanent except change. To emphasize the universality and interconnectedness of change, for everything in Nature seems related to everything else, a descriptive adjective from the Greek *kosmos,* meaning an orderly whole, does seem appropriate. We thus grant this process of "universal change" a more elegant, broad name—cosmic evolution—and we propose it as a majestic worldview that incorporates living beings quite naturally into the larger realm of all material things.

### The Arrow of Time

Consider, as shown in Figure 1, the arrow of time—an archetypal illustration of cosmic evolution. Regardless of its shape or orientation, such an arrow represents an intellectual guide to the *sequence* of events that have changed systems from simplicity to complexity, from inorganic to organic, from chaos to order. That sequence, as determined from a substantial body of post-Renaissance observations, is galaxies first, then stars, followed by planets, and eventually life forms.[1] In particular, we can identify seven major construction phases in the history of the Universe (denoted diagonally in Figure 1): particulate, galactic, stellar, planetary, chemical, biological, and cultural evolution. These are the individual phases, separated by discontinuities from a myopic perspective, that demarcate the disciplinary and fragmented fields of today's specialized sciences.

As such, the most familiar kind of evolution—biological evolution, or neo-Darwinism—is just one subset of a much broader evolutionary scheme encompassing much more than mere life on Earth. In short, what Darwinism does for plants and animals, cosmic evolution aspires to do for all things. And if Darwinism created a revolution in understanding by helping to free us from the anthropocentric notion that humans basically differ from other life forms on our planet, then cosmic evolution is destined to extend that intellectual revolution by releasing us from regarding matter on Earth and in our bodies any differently from that in the stars and galaxies beyond.

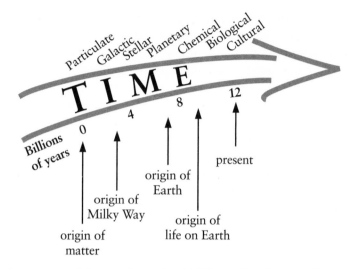

Figure 1. An arrow of time can be used to highlight salient features of cosmic history, from the beginning of the Universe to the present. Sketched diagonally along the top of this arrow are the major evolutionary phases that have acted, in turn, to yield increasing amounts of order and complexity among all material things. Despite its implication of "time marching on," the arrow is meant to imply nothing strictly deterministic, nor progressive. Much as for its most celebrated component—neo-Darwinism—the twin agents of chance and necessity embed all aspects of the cosmic-evolutionary scenario, whose temporal "arrow" hereby represents a convenient guide to natural history's many varied changes.

Of central importance, we can now trace a chain of knowledge—a loose continuity along an impressive hierarchy—linking the evolution of primal energy into elementary particles, the evolution of those particles into atoms, in turn of those atoms into galaxies and stars, the evolution of stars into heavy elements, the evolution of those elements into the molecular building blocks of life, of those molecules into life itself, of advanced life forms into intelligence, and of intelligent life into the cultured and technological civilization that we now share. These are the historical phases, much the same as those noted above but now reidentified from an integrated perspective, that comprise the interdisciplinary worldview of the present work. The attitude here is that, despite the compartmentalization of modern science, evolution knows no disciplinary boundaries.

Broadly conceived in this way, and despite its name, cosmic evolution is not confined to those changes within and among astronomical objects. Rather, this universal subject encompasses all change, on every spatial and temporal domain—large and small, near and far, past and future. Accordingly, neo-Darwinism becomes but one, albeit important, part of an extensive evolutionary scenario stretching across all of space and all of time.

Nor is cosmic evolution an attempt to extrapolate the core Darwinian principle of natural selection to realms beyond life forms. Ambitiously, it is more than that. Cosmic evolution is a search for principles that subsume, and even transcend, Darwinian selection—a unifying law, an underlying pattern, or an ongoing process perhaps, that creates, orders, and maintains all structure in the Universe, in short a search for a principle of cosmic selection.[2]

Metaphorically (at least), cosmic evolution aims to frame a heritage—a cosmic heritage—a grand structure of understanding rooted in events of the past, a sweeping intellectual map embraced by humans of the present, a virtual blueprint for survival along the arrow of time if our descendants of the future are to realize a future. The objective, boldly stated, is nothing less than a holistic cosmology in which life has not merely a place in the Universe, but also perhaps a significant role to play as well.

In effect, with cosmic evolution as the core, we espouse a new philosophy—a scientific philosophy. And we hasten to place due emphasis on that key adjective, "scientific." For unlike classical philosophy, observation and experimentation are vital features of this current effort; neither thought alone nor belief alone will ever make the unknown known. Cosmic evolution strives to address the fundamental and age-old questions that philosophers and theologians have traditionally asked, but to do so using the scientific method and its technological instruments—the most powerful twin techniques ever invented for the advancement of factual information.

Indeed, the same technology that threatens to doom us now stands ready to probe meaningfully some of the most basic questions ever asked: Who are we? Where did we come from? How did everything around us, on Earth and in the heavens, originate? What is the source of order, form, and structure characterizing all material things? How did (and does) order emerge from disorder, given that the second law of thermodynamics dictates the Universe to become increasingly randomized and unstructured? Can we reconcile the theoretical destructiveness of thermodynamics (often called the "thermodynamic arrow of time") with the observed constructiveness of cosmic evolution (the "cosmological arrow of time")? Of ultimate import, armed with a renewed and quantified perception of change, science now seems poised to address the origin of the primal energy at creation itself, and thus to tackle the fundamentally fundamental query: Why is there something rather than nothing?

However time flows and for how long, we take it to be a linear phenomenon, to unfold at a steady pace from its fiery origins to the here and now of the present. Likewise do we invoke the unchanging character of the physical laws (Feynman 1967), for without these assumptions we cannot meaningfully proceed to investigate our ancient past. These are also among the same assumptions underlying most studies of the far future (Dyson 1979), which is not a topic of this book; without unvarying constants of Nature and fixed principles of science, no objective advance can be made in understanding. All this is tantamount to saying that $2 + 2 = 4$ throughout the Universe or that hydrogen atoms are built identically everywhere; if these central tenets are untrue, then read no further.

### No Anthropocentric Agenda

Despite its clean and simple lines in Figure 1, the arrow of time harbors no implication of progression or directedness, no action that unhesitatingly and inevitably leads from the early Universe to ourselves. Anthropocentrism is neither implied nor intended by this arrow; no logic supports the idea that the Universe was conceived (or self-conceived) in order to produce us. We humans are surely not the culmination of the cosmic-evolutionary scenario, nor are we likely to be the only technologically sentient beings that have (or will have) emerged in the organically rich Universe. Flatly stated, there is here no veiled attempt, or hidden agenda, to exalt humankind or to place our species atop some elevated pedestal. As a philosophy of approach, and in keeping with empirical findings, the Copernican principle is in full force throughout this book, denying Sun, Earth, and life any special status—in time, in space, or in complexity.

Nor do we seek to resurrect the *scala natura,* or linear ladder of life, that ancient Aristotelian (or Hsun Ch'ing before him) stepwise hierarchy, or "great chain of being," of bygone pre-evolutionary days. "Lower," primitive organisms do not biologically change into "higher," advanced organisms, any more than galaxies physically change into stars, or stars into planets. Rather, with time—much time—the environmental conditions suitable for spawning primitive life forms eventually gave way to those favoring the emergence of more complex species; likewise, in the earlier Universe, the environment was ripe for galactic formation, but now those conditions are more conducive to stellar and planetary formation. Change in the surrounding environment usually precedes change in an organized system per se, and those system changes have *generally* been toward greater amounts of diversity and complexity. The popular image of a straight and narrow ladder of life itself evolved in Darwin's day into the metaphor of a branching treelike structure, with the simpler, mostly extinct and fossilized life forms comprising the thicker, bottom bulk of the tree, and the more complicated, currently living forms adorning its thinner limbs near the top. Nowadays, it is more common to imagine life's evolutionary model as a scrubby bush, or even a landscape of many stunted bushes amidst tall grasses and perhaps a few trees (Gould 1980).

The arrow of time in Figure 1 provides a simplified context for the rich natural history of all events that preceded us; it has worked well in classroom settings as a clear, compelling symbol. Even so, for those who would have trouble with such an innocent illustration in our cosmic-evolutionary lexicon, I offer Figure 2 as an alternative, perhaps less varnished rendering of events from the beginning of time to the present. If the thin, sleek arrow of Figure 1 is akin to a rather pruned tree of life, then the wider, yet cramped arrow of Figure 2 is analogous to the more realistic bush of life. Here, the dynamic tide of ceaseless change is portrayed in a more contorted fashion, the entire Universe resembling an intricate web of step-by-step causality. All the while, for either arrow, time is assumed to move linearly—granting evolution a partly random, undirected pace while building structures from spiral galaxies to barren planets to reproductive beings—though still sequentially, largely according to the structures' degree of complexity. Onward and upward? No, just onward; we cannot recapitulate enough that cosmic evolution entails no progress or design that equates humankind with the goal of some magisterial plan. Our deep-seated anti-anthropocentrism is one reason (the irreversibility of thermodynamics is another) that we prefer the symbol of a thickening, sideways-flowing arrow to any thinning, upward-thrusting ladder, tree, or bush.

Contingency—randomness, chance, stochasticity—pervades all of dynamic change on every spatial and temporal scale, an issue to which this book returns repeatedly. And with this uncertainty, we emphasize that science today is no longer in the prediction business, at least not nearly as much as in the older, Newtonian worldview; cosmic evolution predicts little of the future, yet strives to explain much of the past.

What about the anthropic principle, a nagging conundrum that just

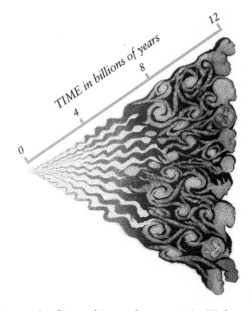

Figure 2. Time's arrow implies nothing anthropocentric. We humans are neither likely the pinnacle, nor surely the end product, of the cosmic evolutionary scenario. The shape of this figure—a more artistic, less diagrammatic "arrow" than in Figure 1—is meant to suggest that the number and diversity of structures have increased with time, yet without any kind of resemblance to a classical evolutionary tree having main trunks and well-defined branches. Whatever measure of complexity is used, it is hard to avoid the notion that "things"— whether galactic clouds, slimy invertebrates, luxury automobiles, or the whole Universe itself—have generally become more complicated throughout the course of history.

won't go away in cosmological circles (Carter 1974; Wald 1984; Barrow and Tipler 1986)? Stated in one of its many versions, "The Universe is the way it is because we are here to observe it." Or in another, "The Universe becomes knowable when there is someone to observe it." Or even, "We are here because the Universe is designed for us." All of these statements court anthropocentrism (as the principle's name implies) and, coincidences aside, a medieval purposive organization of matter—which is why the arguments of this book reject the anthropic principle's strong form, all the while accepting its weak form.

The issue is this: If the numerical values of certain physical constants (for example, the velocity of light, an electron's charge and mass, the gravitational constant, etc.) differed even slightly from their observed values, then the long sequence of events that produced galaxies, stars, planets, and life might have been impossible. The cosmos would likely be starless and lifeless, a proposition very much at odds with the one seen around us. A good example concerns the basic constant of gravity, $G = 6.67 \times 10^{-8}$ dyne-cm$^2$ g$^{-2}$: If G were much smaller, matter would not be able to compress enough to create the temperature and density needed for hydrogen ignition, hence stars would not have formed from dark balls of gas; if G were much larger, stars would have formed but would have burned hotter and endured for less time, making it improbable that life would have originated on any attendant planets. Another example is that if the foremost number in quantum mechanics—Planck's constant, $h = 6.63 \times 10^{-27}$ erg-s—were even a few percent larger or smaller, the nuclear reactions that create carbon in stars wouldn't work. Yet without carbon and the multiple bonding ability it confers on complex chemical structures, life as we know it couldn't exist. These hypotheticals suggest that there is a relatively small window of numerical values that would allow the existence of cosmos, stars, and life.

The "strong anthropic principle" implies that our Universe is very finely tuned—as if by fiat—in order to produce precisely certain kinds of structures that are observed, including, ultimately and inevitably, intelligent life. Those who subscribe to this extreme version seemingly accept an agenda that borders on the mystical, the implication being that the Universe is a goal-oriented, comfortable abode perfectly tailored for the emergence of intelligent life. Proponents apparently want to believe that humanity is exceptional, even perhaps unique, as though the Universe has toiled specifically and necessarily to yield us. However, strongly anthropic reasoning is demonstrably tautological, even teleological—humankind's latest attempt to reinstate for itself a special position in the cosmos, to argue for a purposeful design, and thence a Designer, in Nature.

Multiple universes are often postulated by those troubled by the strong anthropic principle, and who wish to avoid its above-stated dilemma without resorting to unacceptably large coincidences. A whole family of universes, all simultaneously present yet each with a different set of physical constants, would permit one such Universe—namely, the one we inhabit—to have the "right" set of constants for the onset and endurance of stars, life, and all the other complex structures around us. However, the concept of multiple universes is unviable, implying a huge semantics problem. It represents an attractive idea in science fiction, yet without basis in science fact. The Universe is all that there is, by definition: "The totality of all known or supposed objects or phenomena, formerly existing, now present, or to come, taken as a whole." An inability to observe other universes, or even to test experimentally for their existence, will indefinitely keep this bizarre idea safely outside the realm of science.

The concept of an oscillating Universe might also enable us to deny the issue of purpose seemingly injected into cosmology by the strong anthropic principle. Here, the cosmos repeatedly expands and contracts in a series of "multi-verses" that exist one at a time, all of them together having neither beginning nor end, yet each cycle of oscillation endowed with a different set of physical constants. Accordingly, we would now happen to live in that cycle for which those constants are conducive to life and intelligence. Alas, there is not a shred of observational evidence to favor an oscillating Universe, however appealing such a big-bang-less model would be philosophically.

In contrast to the strong version, the "weak anthropic principle" is more palatable, indeed should present us no trouble since it hardly more than restates cosmic evolution itself. The weak version acknowledges that the Universe does seem well tuned, yet suggests that Nature's basic forces just happen to have the right strengths to evolve hierarchically ordered structures over long periods of time. This is not a coincidence. Rather, those structures, including life, adapt to the prevailing physical conditions, whatever they may be, eventually filling an entire spectrum of environmental niches with a variety of ordered types, in fact those very galaxies, stars, planets, and life forms enveloping us in today's observable Universe. Furthermore, the weak principle implies that if one or another physical constant differs from its known value then others among those constants would also likely differ—probably in such a way as to permit the contingency-oriented construction of ordered structures representing a statistically indistinguishable set of galaxies, stars, planets, and again life. The precise values of the physical constants measured today are not necessarily the only combination of values that could conceivably lead, through successive hierarchical stages discussed in this book, to intelligent life itself. No part of the anthropic principle explains *why* the physical constants have their present values. Tackling *why* questions is not currently part of the fabric of science. By contrast, addressing *what* questions (the inventory) and *how* questions (the dynamic) are the sum and substance of mainstream science.

The weak anthropic principle also entails the idea that life can arise and exist only during a certain epoch in time—and that humankind now happens to populate that epoch. But that again is nothing more than an expression of cosmic evolution naturally unfolding and should not endow us with any speciality. That is, life (as we know it) could arise only after sufficient amounts of carbon were made in stars and would thereafter likely terminate when all the stars themselves burned out. This well-defined (and long!) time interval is completely consistent with cosmic evolution as discussed in this book, without any obligation to regard ourselves—or any life forms, here or elsewhere—as privileged. Cosmic evolution grants us a materialistic avenue to explore the origins of matter and life without undue romanticism or flights of fancy among those who prefer to anoint life, and especially humankind, with the status of *sui generis*.

Hubert Reeves, the French astrophysicist (1991), has it right when, rather than embracing the chauvinistic term "anthropic principle" in any of its sundry versions, he prefers to adopt the expression, "complexity principle," for all organized entities, to wit: "Since the earliest times accessible to our exploration, the Universe has possessed the properties required to enable matter to ascend the pyramid of complexity." To be sure, complexity is a, and perhaps *the,* key to both deep and broad understanding in the natural sciences, not only to show that the wonder and delight for our natural cosmos is comprehensible, but specifically to show how the cosmos can be comprehended without destroying the wonder. For when we have deciphered the underlying pattern and have explained the wonderful in Nature, a new wonder will arise at how complexity has been forged out of simplicity, conscious beings out of primitive elements.

## Rise of Complexity

Cosmic evolution accords well with observations revealing that an entire hierarchy of localized structures has become enriched, in turn, during the history of the Universe: particles, atoms, galaxies, stars, planets, life, intelligence, and culture. Figure 3 graphs this widespread impression that material assemblages have become increasingly organized and complex, especially in relatively recent times—ordered systems of much variety, diversity, and intricacy in an otherwise disordered and chaotic environment. Nothing specific or quantitative is implied by this graph; it is drawn here merely to sketch the widely ac-

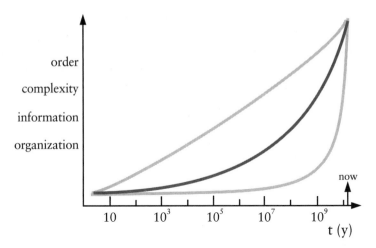

Figure 3. Sketched here qualitatively is the rise of order, form, complexity, organization, inhomogeneity, clumpiness, structure, information—or whatever broadly synonymous term one wants to use to describe the development and evolution of *localized* material assemblages—throughout the history of the Universe. (Note that the horizontal, temporal scale is logarithmic, in contrast to the linear scales of the first two figures, a change better able to illustrate certain details as this book unfolds, and one consistent with all subsequent graphs; time is still assumed to flow linearly.) That the complexity of ordered structures has generally risen over the course of time is well-recognized, albeit difficult to measure; the intermediate, darker curve drawn here represents an innate feeling for steeply rising complexity in more recent times, although that rise might have been either more gradual (even linear) or more exponential (even hyperbolic) as also drawn in lighter grey. Our main task in this book is to explain the *general trend* of these curves, and to do so in quantitative terms to the extent possible, with known, modern scientific principles. (For a preview of things to come, compare with Figure 28.)

cepted idea that "islands" of complexity have emerged over the course of time, and, furthermore, dramatically so during the past 500 million years, or roughly a few percent of the age of the Universe. There are more than a hundred kinds of atoms (physical phenomena), hundreds of thousands of types of organic molecules (biochemical phenomena), and millions of species of living organisms (biological phenomena). My goal in this book is to explain the general trend of rising complexity represented by the family of curves of Figure 3. We cannot predict where it might be headed in the future, but we can describe how it came to be in the past.

By "complexity," we refer to the term intuitively as used in ordinary discourse, a definition culled from many sources: "a state of intricacy, complication, variety, or involvement, as in the interconnected parts of a structure—a quality of having many interacting, different components." In this work, we shall come to identify complexity in two operational ways: as a measure of the information needed to describe a system's structure and function, or as a measure of the rate of energy flowing through a system of given mass. No attempt is made here to be rigorous with the words "order," "organization," "complexity," and the like; this is not a work of classical philology or linguistic gymnastics. Indeed, no two researchers seem able to agree on a precise, technical definition of such a specious word as complexity, which may be context-dependent in any case. (For a discussion of the subtle differences among these approximately synonymous terms, see Davies 1988, Csányi 1989, Lewin 1992, or Corning 1998.) In particular, note that we are not merely appealing to the so-called "complex adaptive systems" often used to model intelligent interacting agents (like "smart" behavioral or social systems that display learning) in the newly emerging complexity sciences (Gell-Mann 1994). Although the localized systems described herein are too complex to be described with the two-body science of old, their sufficiently large number of (often "dumb") interacting agents do lend themselves to statistical study of their net dynamics and broad makeup.

What this book seeks to explore, in liberal thermodynamic terms, is the way in which any system's components, apart from its environment, act or are arranged in a cooperative, systematic fashion. Our emphasis here is on structural, or morphological, complexity—both internal and external anatomy—and less on functional, or behavioral, complexity. Such a trend toward increasing complexity in biological evolution was asserted more than a century ago by geologist Louis Dollo (1893) and philosopher Herbert Spencer (1896), an intuition reinforced since by many others (see, for example, Jantsch 1980, Bonner 1988, and Kauffman 1991). Similar ideas were, in turn, extended beyond biology by fewer scholars over the years, notably the French positivist Auguste Comte (1842), who proposed a sequence of decreasing "abstractions" uniting all the sciences from mathematics to astronomy to physics to chemistry to biology and on to psychology, and the American astronomer Harlow Shapley (1930), whose "cosmography," or comprehensive classification akin to zoology, ordered all known structures by increasing dimensions, including atoms, molecules, planets, stars, galaxies, and not least "higher combinations of sidereal systems." Later in his 1967 memoirs, Shapley—the other of my intellectual heroes—went beyond mere classification, qualitatively alluding to an evolutionary link throughout all of Nature: "nothing seems to be more important philosophically than the revelation that the evolutionary drive, which has in recent years swept over the whole field of biology, also includes in its sweep the evolution of galaxies, and stars, and comets, and atoms, and indeed all things material." His vision closely parallels the quantitative analysis of the work presented here.

Admittedly, disorder occasionally gains the upper hand at isolated points in space and time, such as when a supernova detonates, thereby destroying a star's thermal and chemical order concentrically fashioned over eons; or when a bat-like creature retreats over generations into a cave, thereby gradually becoming "simpler" by, for example, losing its eyesight (as it then has no evolutionary advantage); or even when a highly stratified social system such as the Maya declines, thereby reducing its cultural complexity while biological individuals survive. Mass extinctions, as documented in the fossil record, represent additional examples of departures from an otherwise impressive evolution of life toward more complex forms. But such decreases in complexity are comparatively rare, implying that cosmic evolution is not some kind of random walk; the general trend is demonstrably toward increased complexity with increased time. And in this book, we address the universal, long-term tendencies, the bigger picture, thus recognizing an overall rise in complexity with the inexorable march of time—a distinctly temporalized "cosmic change of being," without any notion of progress, purpose, or design implied.

With cosmic evolution as an intellectual framework, we can begin to understand the environmental conditions needed for material assemblages to have become increasingly ordered, organized, and intricately structured, and not merely among biological systems. The rise in order, form, and complexity applies to all animate and inanimate organized systems; the trend represented by the curves of Figure 3 violates no laws of physics, and certainly not those of modern thermodynamics. Indeed it is modern thermodynamics that perhaps best helps to explain it. Nor is the idea of ubiquitous change novel to our contemporary knowledge of the world, the Universe, and ourselves. What is new and exciting—and this is the main thrust of this short book—is the way that frontier, non-equilibrium science now helps us mold a holistic cosmology wherein life does play an integral role.

## Summary of Cosmic Evolution

This book surveys the grand scenario of cosmic evolution by qualitatively and quantitatively examining natural changes among radiation, matter, and life within the context of big-bang cosmology. The early Universe is shown to have been flooded with pure energy whose radiation energy density was initially so high as to preclude the existence of any appreciable structures. As the Universe cooled and thinned, a preeminent phase change occurred about a hundred millennia after the origin of all things, by which time matter's energy density had overthrown the earlier primacy of radiation. Only with the onset of technologically manipulative beings (on Earth and perhaps elsewhere) some 12 billion years later has the energy density contained within matter become, in turn, locally dominated by the rate of free energy density flowing through open organic structures.

Nature has, surely naturally and perhaps inevitably, managed to establish the environmental conditions conducive to the complexification of all known systems within an otherwise disordered Universe. Using non-equilibrium thermodynamics as the crux, I argue that it is the contrasting temporal behavior of various energy densities that has given rise to those environments needed for the emergence of galaxies, stars, planets, and life. Energy flow diagnostics are seen to be especially useful in tracking through time the rise of complexity among a variety of open systems, animate and inanimate. I furthermore maintain that a necessary, though perhaps not sufficient, condition—a veritable prime mover—for the onset of such ordered structures is the expansion of the Universe itself. Neither manifestly new science nor appeals to non-science are needed to understand the impressive hierarchy of the cosmic-evolutionary scenario, from quark to quasar, from microbe to mind.

Cosmic evolution is the study of many varied changes on a universal scale, a subject that seeks to synthesize the reductionist posture of specialized natural science with a holistic view that goes well beyond. It is a story about the awe and majesty of twirling galaxies and shining stars, of redwood trees and buzzing bees, of a Universe that has come to know itself. But it is also a story about our human selves—our origin, our existence, and perhaps our destiny.

# THE NATURE OF CHANGE

C osmic evolution, as we understand it today, is governed largely by the laws of physics, particularly those of thermodynamics. After all, of all the known principles of Nature, thermodynamics has perhaps the most to say about the concept of change. Literally, thermodynamics means "movement of heat"; a more insightful translation (in keeping with the wider connotation in Greek antiquity of motion as change) would be "change of energy."

## Thermodynamics

The first law of thermodynamics is a conservation principle. It states that all energy in the Universe is constant—that is, the sum of all energy is fixed, has been fixed since the beginning of time, and will remain so until the end of time. Formulated independently in the 1840s by two German medical doctors, Robert Mayer and Hermann von Helmholtz, this law has no known exceptions to date. Even so, energy can appear in various forms, for example, heat, light, gravitation, invisible radiation, kinetic energy, mechanical work, chemical potential, nuclear energy, and so forth; matter itself is a form of energy. Furthermore, the many varied forms of energy can be interchanged, including matter transforming into energy and conversely—as Albert Einstein taught us in 1905—a fact made evident by the development of atomic weapons and nuclear reactors. In short, the first law of thermodynamics decrees that energy itself can be neither created nor destroyed, though its many forms can change.

If the first law were the totality of thermodynamics, we could interchange energy among its varied forms (including matter) without limit. Alas, there exists another basic principle of thermodynamics, a second law, which is more subtle than the first. Scientifically expressed in the 1850s by physicists Lord Kelvin of Britain and Rudolf Clausius of Germany independently, the second law of thermodynamics specifies the way in which change occurs in a quantity called "available energy," also occasionally termed "usable energy," "free energy," or "potential energy." This law's essence stipulates that a price is paid each time energy changes from one form to another. The price paid (to Nature) is a loss in the amount of available energy capable of performing work of some kind in the future. In physics, entropy is the term characterizing this decrease in available energy; it derives from a Greek root meaning "turn" or "change." Clausius himself, in 1865, explained the origin of this peculiar name: "This term is based on the Greek word *tropae*, meaning 'transformation.' I have deliberately made the structure of this word analogous to that of 'energy,' because the two quantities described by these terms [energy and entropy] are so closely related in their physical meanings that the parallel designation seems useful here."

Entropy, when multiplied by temperature, is a measure of the amount of energy no longer capable of conversion into useful work. Numerically, the change in entropy, $\delta S$, varies directly as the heat exchanged, $\delta Q$, between two systems ("heat" being the name of the transfer process itself, that is, the amount of energy in transit), and indirectly as the temperature, T, at which the change occurs, namely,

$$\delta S = \delta Q / T .$$

Entropy is also a measure of the disorder (or randomness) of a system, whether that system is something as small as a crystal of molecules (or even smaller) or as large as a cluster of galaxies (or even larger).

Actually, some of these basic thermodynamic ideas date back even further, to 1824, when a young French army officer, Sadi Carnot, sought to understand the rudiments of an ordinary steam engine. He discovered that such engines work because of a temperature difference; part of the engine is cold while another part is hot. Indeed, for any engine to convert energy into useful work, there must be a temperature differential (or thermal gradient). Work can then occur when energy (heat in the case of a steam engine) flows from a body of higher temperature to one of lower temperature—in other words, from a higher energy state to a lower energy state. Carnot articulated his findings succinctly in his memoirs, *Reflections on the Motive Power of Fire:* "For a heat engine operating in cycles to perform mechanical work, we must use two bodies of different temperatures." Equally important, Carnot discovered that each time energy flows from one state to another, less energy is available to perform work the next time around. Energy is not lost, just rendered unavailable for useful work.

Thus, if any isolated physical system—one completely disconnected from the outside world—is divided into two parts, energy can flow from one part to the other, but the total energy of the system cannot be increased or decreased. This, in essence, is a statement of the first law of thermodynamics. Accompanying this internal flow of energy will also be a flow (or change) of entropy. Only for ideal (so-called reversible) processes will the resulting change in the entropy of the system be zero. For realistic (i.e., irreversible) processes, there must be an increase in the *total* entropy, for with each such process less energy is available to do something useful. This, in turn, is a statement of the second law of thermodynamics. In symbolic form, $E = F + TS$, and stressing differences as we must (hence the $\delta$) for the hallmark of any process is the *change* in a system's state,

$$\delta E = \delta F + T\delta S + S\delta T \, ,$$

where E is the total internal energy of the system and F is its free, or available, energy capable of doing work. (F is often technically known among physicists as the Helmholtz free energy, in contrast to the slightly different Gibbs free energy more commonly used by chemists. Conceptually their roles are similar; their numerical differences often small, in fact inconsequential for biological systems.) Hence, for fixed energy and a given temperature ($\delta T = 0$), if F decreases then S must increase. Figure 4 captures these ideas graphically and Figure 5 gives a numerical example.

What is the relative importance of the two laws? The British astronomer, Sir Arthur Eddington, answered rather directly during the 1927 Gifford Lectures: "The law that entropy always increases—the Second

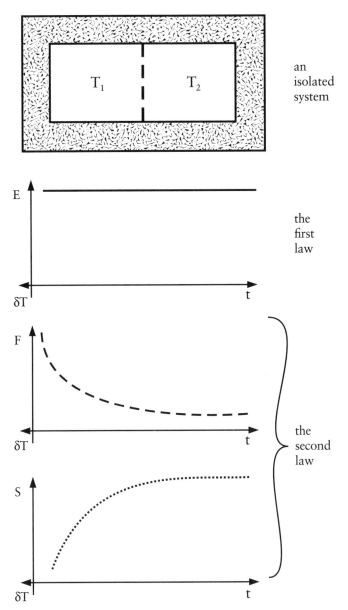

an
isolated
system

the
first
law

the
second
law

Figure 4. The essence of classical, irreversible thermodynamics: In a system isolated from the outside world, heat within a gas of temperature, $T_2$, will flow in time, t, toward a gas of temperature, $T_1$, where $T_2 > T_1$ and $\delta T = T_2 - T_1$, thus conserving the system's total energy, E (via the first law of thermodynamics), all the while its free energy, F, decreases and its entropy, S, rises (via the second law of thermodynamics).

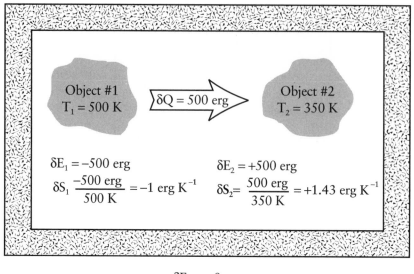

$$\delta E_{total} = 0 \text{ erg}$$
$$\delta S_{total} = +0.43 \text{ erg K}^{-1}$$

Figure 5. Energy and entropy changes computed for the case of an isolated system wherein 500 ergs of thermal energy ($\delta Q$) flow from a 500 K object to a 350 K object. The amount of heat transferred is so small that it does not affect the temperature of either object. Net energy is conserved, $\delta E_{total} = 0$ erg, in accord with the first law of thermodynamics; net entropy increases, $\delta S_{total} = 0.43$ erg K$^{-1}$, in accord with the second law. In short, the entropy increase of the cooler object exceeds the entropy decrease of the warmer object.

Law of Thermodynamics—holds, I think, the supreme position among the laws of Nature. If someone points out to you that your pet theory of the Universe is in disagreement with Maxwell's equations—then so much the worse for Maxwell's equations. If it is found to be contradicted by observation—well, these experimentalists do bungle things sometimes. But if your theory is found to be against the Second Law of Thermodynamics, I can give you no hope; there is nothing for it but to collapse in deepest humiliation."

A familiar example provides an especially good illustration of the above ideas. When an iron bar is heated at one end, the other end will eventually warm until the temperature of the whole bar becomes equal. This is known to anyone who has left for too long a metal stirring spoon in a pot of hot soup or a poker in a fireplace. The reverse phenomenon—namely, a uniformly warm iron bar suddenly becoming hot

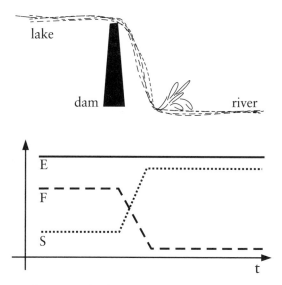

Figure 6. As water flows over the dam, gravitational potential energy (or free energy, F, dashed line) can be converted, by means of a water wheel or turbine, into useful mechanical or electrical energy. In the process, as time, t, advances, F decreases (mimicking the water's drop), the system's total energy, E (solid line), remains the same, and the entropy, S (dotted line) increases.

at one end and cold at the other—has never been observed. Likewise, when a container partitioned into two compartments, one filled with gas and other empty, has its partition removed, the gas will expand into the empty compartment until both parts are equally full. The reverse process—namely, the dispersed gas abruptly congregating in only one half of the container—has also never been observed. (Have you ever seen the steam released from a whistling teakettle suddenly heading back into the kettle?) All such natural (realistic) events proceed in one direction only. Nature forbids (or at least demands a penalty for) their reversal, which is why real events are described as irreversible or asymmetric: Hot objects cool, but cool objects do not spontaneously become hot; a bouncing ball comes to rest, but a stationary ball does not spontaneously begin bouncing.

Consider yet another familiar example of irreversible energy change—namely the case of water falling over a dam into a basin below, as sketched in Figure 6. This is a mechanical (or gravitational) analog of the thermal and chemical cases just discussed and is the kind of energy that powered textile mills during the Industrial Revolution.

"Spindle cities," such as Manchester, England, and Lowell, Massachusetts, developed adjacent to rivers whose waters were diverted into an intricate system of canals to provide the primary source of energy to turn huge waterwheels mechanically. Later, about a century ago (and partly continuing to the present), these same systems powered turbines that generated much of the electricity for those cities. Many of their man-made canals utilized locks, or artificial dams, to hold the water temporarily at certain levels, after which the water could experience a drop in level and thus power the weaving machines that once manufactured textiles. Having traversed the canal system and reached the river basin below, however, the water could no longer perform work. Still water in a dammed river or lake cannot be used to turn even the smallest turbine. Nor can the basin's water be returned to the top of the dam without work being performed by some outside agent (such as a water pump), for this realistic example is an irreversible process. (Of course, as agents of change, we could use just such a pump, but this artificial process is decidedly uneconomical; more energy is required than is produced.) In short, water in a reservoir above a dam possesses some available or free energy—something we often call "gravitational potential energy," which is yet another form of energy that can be released as the water falls to a lower level closer to the center of the Earth. Provided that an energy difference exists (in this case, gravitationally owing to differing heights of the water), then useful work can occur. In the basin below, the water still harbors energy (say, of a thermal or turbulent nature), but it is unavailable for useful work since the energy of the water in a flat basin is everywhere the same. Furthermore, each time more energy is made unavailable for work (in this case by falling over the dam), entropy increases. The entrepreneurs who backed the early industrialists some two centuries ago were essentially performing a controlled experiment in gravitational contraction of the Earth—a much more efficient way to produce energy than in a chemical reaction or nuclear power plant.

So, the first law states that energy can be neither created nor destroyed, while the second law stipulates that energy can change in only one way: irreversibly toward a dissipated (randomized) state of increased entropy. Nature is said to be intrinsically asymmetric. Furthermore, nothing is inevitable about the outcome of the two laws; thermodynamics places limits on what is possible, telling us if a process can occur, not whether it will occur. Nature is also capricious, an issue to which we shall return, and expand upon, several times in this book.

## Equilibrium

Let us again restrict ourselves to an isolated system, that is, one cut off from its surrounding environment and into which no new energy or matter flows. In such a system, energy states always tend to even out, that is, achieve an equilibrium. In all the examples cited above—heat spreading along a metal bar, gas diffusing within a bounded container, water flowing into a river basin—equilibrium is the end product. That is the meaning behind the second law of thermodynamics: Nature harbors an inherent tendency to eliminate inequalities and realize uniformities in the distribution of matter and energy, to achieve a maximum-entropy configuration for any unconstrained system. The following cases further serve to illustrate operationally the nature of an equilibrium.

A balance scale maintains equilibrium when both its pans hold equal weights. If we tap the beam, the scale will oscillate, as we have imparted to the system some potential energy that converts into kinetic energy and then back into potential energy and so on. The total energy of the system remains constant, and the scale would oscillate indefinitely if it were not for friction. Realistically, however, friction will eventually cause the oscillations to diminish, and the energy contained in the (directed) oscillatory motions will become changed into the (undirected) heat motions of the individual atoms comprising the scale—all the while increasing its entropy while striving to achieve that state of equilibrium typified by the motionless scale in complete balance. Likewise, a pendulum achieves momentary equilibrium each time its arc is centered in the middle of its swing. (Even the "balance of power" characterizing the Cold War's international politics was an equilibrium of sorts.)

Burning wood in a stove or fireplace is yet another example of irreversible change toward equilibrium. The fire causes heat (i.e., infrared energy) to radiate and thus warm a room, as the second law stipulates that heat always flows from the hotter body (the stove) to the colder body (the surrounding air). Eventually, though, the wood will have burned completely and its remaining ashes will have reached the same temperature as the air in the room. Whereas previously there was a distinct difference in energy states with the fire burning, day-old ashes exhibit no differences in energy states and have thus achieved an equilibrium. Like water in a flat basin or a pendulum at rest, spent wood is no longer capable of performing useful work on its own. Again, additional water could be hoisted to the top of the dam and additional wood could be placed into the hearth, but these actions would involve the use of a new source of available or free energy and thus would violate our restriction of an isolated system.

The equilibrium state achieved within an isolated system is therefore a condition of maximum entropy—a stable state where energy can no longer freely perform useful work. In short, an equilibrium is characterized by an absolute minimum of free energy and a consequent maximization of entropy. Interestingly enough, only in equilibrium can we not distinguish past from future, for in such a state change has no direction; the concept of irreversibility itself ceases at equilibrium.

### Order and Disorder

Recall that we earlier equated entropy with randomness; the greater the randomness or disorder, the greater the entropy. To see this, we can regard the observable macroscopic properties of any system as the sum of a great many microscopic properties—perhaps the energies of electrons in an atom, or the vibrations of atoms in a molecule, or the motions of molecules in a gas. Any system having great randomness—that is, high entropy—is one having these and other microscopic properties arranged in many different ways. Conversely, a system of low entropy has only a few possible arrangements of its inherent properties.

A familiar analogy of entropy, randomness, and disorder is a cluttered dormitory room. There, books, pencils, socks, or scarfs, among scores of other personal items, might be almost anywhere—on a shelf, windowsill, chair, bed, floor, wherever. Myriad arrangements are possible in a messy room, compared to only one or a few such arrangements when the room is clean, that is, when all the items are in their proper places. Although only an analogy and not an entirely accurate translation of a subtle statistical concept, such a familiar illustration allows us to glimpse how disorder can be associated with a large number of different possibilities—or spatial configurations—among the many sundry items comprising a single room.

Another telling illustration is the state of any university library at the end of each semester. Some of us in science hold high the need for clear communication and hence require students to write term papers in many of the courses we teach. This inevitably leads to considerable confusion in the library, for it is a common shortcoming of college life that students (and faculty, too) cannot seem to reshelve library books properly according to their call numbers. Thus, as students research their topics toward the end of the term, the books are constantly coming off the shelves but more often than not are incorrectly replaced. And of course, if a library book is improperly shelved, it is nearly as good as lost. When the term ends and the library is scanned from a distance ("macroscopically"), all seems fine; few books are missing and the shelves are full. Only upon careful inspection ("microscopically") do we find that chaos rules, and a considerable amount of work (from a kindly librarian) is needed to regenerate order among the volumes. The natural use of a library causes order to break down into disorder; if not periodically checked, entropy will tend toward a maximum in any system.

As a further example, imagine an utterly frozen crystal at absolute zero temperature. Ignoring quantum fluctuations, such a system can have only one possible configuration for its many molecular parts; thus, its entropy is zero, its order high. On the other hand, a gas at ordinary (room) temperature displays much randomness in the distribution of its atoms and molecules; it is said to be disorderly, to have high entropy.

Accordingly, we can also view thermodynamics' second law in another, more profound way. In addition to the notion of energy flowing from available to unavailable states or from high to low concentrations (or temperatures), we can alternatively regard energy in an isolated system as flowing from ordered to disordered states. Minimum entropy states, where energy concentrations are high and available energy is maximized, are considered ordered states. By contrast, maximum entropy states, where available energy is more dissipated and diffused, are considered disordered states.

Thermodynamically, then, order or organization is measured according to the number of possible arrangements of a system's many parts. If a system can be described in terms of only a few such arrangements, we say that the system is orderly. Conversely, if these parts enjoy great freedom in their arrangement, so that they can be described only in terms of many possible arrangements, then we say that the system is highly random or disorderly. And we know that the degree of randomness determines the entropy.

This is the way the second law works in Nature. Left to itself, nothing will proceed spontaneously toward a more ordered state. Housework is a familiar example. Left unattended, houses grow more disorderly; lawns become underbrush, kitchens greasy, roofs leaky. Even human beings who fail to eat will gradually become less ordered and die; and when we die we decay further into a state of even greater disorder, ultimately returning our resources to the Earth and Universe that gave us life. All things, when left alone, eventually degenerate into chaotic, randomized, less ordered states.

Of course, by expending energy, order can be achieved. Some human sweat and hard work—an energy flow—can put a disarrayed house back in order. Recognize, though, when a house is so reordered, it is at the expense of the increasing disorder of the humans cleaning the house; after all, tidying a household is a tiresome activity (especially when a house is populated by curious and exploratory children), often making us feel listless for want of energy. In turn, we humans can become reinvigorated (i.e., personally energized or ordered) by eating food—also an energy flow—but this renewed order is, in turn, secured at the expense of the solar energy that helped initially produce the food. We shall shortly return to discuss more about energy consumption as a means of ordering, reordering, and maintaining complex systems. For now, let us note that this emergence of order from disorder is not a violation of the second law of thermodynamics. This much we now state qualitatively; later we shall prove it quantitatively.

## Open Systems

Earth as a whole has a stock of natural energy resources. Most of the familiar ones—oil, gas, coal—are renewable over geological time scales of millions of years, but must be realistically treated as nonrenewable for our lifetimes and even for the millennial time scales of civilizations. If terrestrial resources were the only sources of energy available on Earth, and no sources existed outside our planet—an isolated Earth system—then, when these sources were completely depleted, differences in energy states would dissipate and Earth would eventually reach equilibrium; all parts of Earth would then attain the same temperature, at which point Earth's available (or free) energy would have zeroed, its entropy maximized.

In point of fact, Earth itself is not an isolated system, nor has our planet yet reached equilibrium; Earth is said to be an "open" system. Our world is still in a formative, albeit prolonged, stage of development. Earth's iron core continues to grow as the planet steadily cools, and its continents collide as its internal energy drives the crustal movements—a good thing, for otherwise the land would be eroded to ever lower elevations, eventually all of it submerged below the waves. Additional, external energy, especially sunlight, reaches Earth's surface daily. Though essentially fixed in rate and pattern of arrival at Earth, solar energy is for all practical purposes unlimited. Long after terrestrial stocks are exhausted, our Sun will still act as a powerful source of energy, a cosmic hearth pouring forth heat and light, thus prohibiting Earth from reaching a state of equilibrium. As diagrammed in Figure 7, the thermodynamics of open systems allow a system's entropy to remain constant or even to decrease. Here, then, is the gist of nonequilibrium thermodynamics, a subject to which we shall return many times throughout the rest of this book: Localized, open systems can be sites of emergent order within a global (i.e., universal) environment that is largely and increasingly disordered. An old adage might draw a helpful distinction: content (the system) is often best judged in context (the environment).

Over very long time scales, the Sun itself will someday run out of fuel, as even today it degrades its own energy supply with each passing second. Without some other source of additional matter or energy, the Sun will eventually equilibrate, the temperature of its matter becoming everywhere the same and the entropy of the entire Solar System maximized. Although nuclear burning in the Sun is scheduled to cease some five billion years hence, thereby effectively equilibrating Earth and maximizing its entropy, such a burned-out star will probably not reach a genuine equilibrium itself for many tens, even hundreds, of billions of years thereafter. Truly great amounts of time are needed for a dead star to pass through the white-dwarf, red-dwarf, and brown-dwarf stages of stellar evolution, eventually becoming a black dwarf— a dark, decrepit, fully equilibrated clinker in space.

To be sure, the second law of thermodynamics has universal applicability; there is no reason to suspect otherwise. Not merely dictating the evolution of Earth and Sun, the second law is presumed to apply also to stars, galaxies, and galaxy clusters, indeed to the Universe as a whole, much like we assume, as nearly an absolute truth as our uncertain age will allow, that energy conservation—the first law—pertains to the cosmos in toto. The Universe currently races outward from the initial big-bang explosion, cooling, thinning, and inevitably developing ever-smaller differences among the energy states of its highly diverse contents. This eventual state of maximum entropy evokes the famous nineteenth-century concept of a "heat death," for the cosmos will have completely run down when all of its energy has been uniformly distrib-

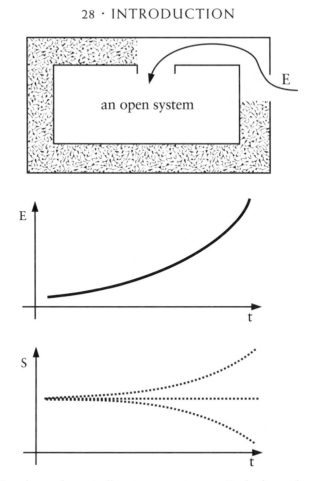

Figure 7. In a thermodynamically open system, energy (in the form of radiation or matter) can enter the system from the outside environment, thereby increasing the system's total energy, E (or more likely its energy density), over the course of time, t. Such energy flow can lead to an increase, a decrease, or no net change at all in the entropy, S, of the system. Even so, the net entropy of the system *and its environment* still does increase, an important point discussed, and proved for some cases, later in this Introduction.

uted; at some time in the exceedingly distant future all available energy in such a Universe would be expended, thus rendering further activity impossible (although by human physiological standards such a depreciated state of zero energy flow would more likely resemble an uncomfortable "cold death"). All galaxies, stars, planets, and life forms will have decayed to their most elementary configurations—even the basic atomic motions will have ceased—as the ultimate equilibrium is achieved. Clausius called it an "unchangeable death," a state of maximum disorder in which all organization and structure will have fully disintegrated and life itself will no longer be possible. As such, cosmic evolution itself would cease, for this is truly a state of eternal rest.

Somewhat by contrast, as discussed later in Chapter Three (Figure 27), a more modern analysis is not so dire, suggesting that the maximum possible entropy will likely never be attained. In an expanding Universe, the actual and maximum entropies both increase, yet not at the same rate; a gap opens between them and grows larger over the course of time, causing the Universe to increasingly depart from Clausius's idealized heat-death scenario. We need not be so pessimistic, indeed it is this inability of the cosmos to ever reach true maximum disorder that allows order, or lack of disorder, to emerge in localized, open systems.

### Time Revisited

The notion of time is basic to a rational inquiry of cosmic evolution, now as well as in the past. Aristotle, among many Greeks of antiquity, regarded time as an absolutely fundamental concept. Mimicking Heraclitus who taught that the Universe is the totality of intangible *events,* not of palpable *things,* Aristotle believed that temporal flux is an intrinsic feature of Nature—but only inside the orbit of the Moon, including events on Earth; all beyond, he thought oddly and wrongly, was fixed and immutable. More than two millennia later, the idea was still popular, at least philosophically, as in, for example, Alfred North Whitehead's notion of "scientific materialism" (a 1925 prelude to his metaphysical organic philosophy), which consisted of nothing else but matter in motion, or a flux of physical energy. As for modern science, it was none other than Einstein (1922) who proposed that time should not be relegated to a poor second; time deserves at least equal footing with the concept of space, thus forming a "spacetime" continuum. That extra dimension is perhaps what best defines the elusive concept of time: the fourth dimension that distinguishes past, present, and future.

But what makes time, an entity that cannot be touched, smelled, or transformed, as much a fiber of our Universe as hydrogen matter or pure energy itself? Is the issue of time so basic to our thinking that we are in danger of overlooking some major assumptions in our discussions of evolution? And is our perception of the direction and rate of time well-founded or merely arbitrary? Not least, what about Einstein's relativistic proviso that time depends on the motion of the observer and the gravitational field in which the observer is located?

An answer to the last of these questions holds that a universal, common time can be invoked to describe the behavior of the whole Universe, even though the Universe is itself in motion and changing its gravitational field. To do so, we appeal to the well-confirmed idea that, on the largest scales (comparable to galaxy superclusters), the Universe looks the same (isotropically) and changes the same (expansively) everywhere; this is the so-called cosmological principle, to be discussed in context in Chapter One. Clocks associated with the Universe en masse—not ones in rapid, relative motion or near intense, localized fields—can then measure time against the grandest of all reference frames defined by the receding, omnipresent galaxies. That is the common, cosmic time used to date all evolutionary events throughout this book—a general advance of space and time in which we observers on Earth participate.

The concept of time is indeed essential to the subject of evolution and especially to the laws of thermodynamics that govern much of it. Unlike events in classical Newtonian physics, which are time-independent, reversible, and ahistorical, in Darwinism the past history of a system contributes to its subsequent properties. As one might expect, the most elementary notion of change cannot be understood without an analysis of time. In our technologically oriented (largely Western) worldview, time "flows" in only one direction—forward. Alternately stated, time is irreversible; "time marches on." Expressed still another way, time moves in whatever way entropy increases; entropy was less in the past and most assuredly will be greater in the future. Eddington put it succinctly: "Entropy is time's arrow."

(We must be careful, since the fact that galaxies, stars, planets, and life forms locally decrease entropy does not mean that their time is reversed. We cannot merely examine a system at two instants of time, determine the time at which the entropy is greater, and then conclude that that is the later instant [Grünbaum 1967]. The larger environments surrounding these objects—realistic, open systems—must be duly considered. And the net entropy change for any open system plus its environment is always positive, thus time transpires without any known reversal.)

The second law of thermodynamics, then, points the direction of temporal change, although it gives no indication of the speed at which change occurs. Nor does it elucidate much about that moment we subjectively call "now," which moves inexorably into the future. According to biologist Harold Blum (1968), the second law "is time's arrow, not time's measuring stick." Nor is the flow of events along time's arrow likely to be constant or uniform; sudden events—supernova explosions, the swat of a fly, collisions among elementary particles—proceed more rapidly than other, slower events in the direction of greater randomness. Thus, although increasing disorder can be generally taken as a measure of the *direction* of time, it cannot be taken as a measure of the *rate* of time's passage. The rate of change of entropy differs for each specific event, all the while the rate of change of time itself is taken (in this book at least) to be uniform, ceaseless, and non-negotiable.

In Figure 1, positioning the salient phases of cosmic evolution precisely on the diagonal astride the arrow of time is not crucial, nor is its left-to-right orientation significant. This symbolic arrow merely provides a wide cognitive map across the evolutionary landscape of major events in the history of the Universe. Time could be visualized equally well moving abstractly from right to left, or up and down, or even in more distorted fashion such as the shape of a knotted, open-ended pretzel. Provided the sequence of events accords with observation—in the main, galaxies originated first, followed by stars, then planets, and eventually life forms (see again note 1 in the Prologue)—the arrow can be drawn with arbitrary shape and orientation. Furthermore, this modeled arrow should be imagined to be flexible, permitting adjustments of the timing of historical events (as new knowledge accumulates) without upsetting the principal temporal sequence. As such, it hardly matters if the Universe is as young as ten billion years or as old as twenty billion years, a controversy encountered at the end of Chapter One. Accordion-like modifications to the arrow of time can expand or contract our description of cosmic history based on the latest data (provided certain constraints are met, such as all objects within the Universe being younger than the Universe *per se*) while still preserving the sequential ordering of cosmic organization, as amply delineated in Figure 1.

To many researchers, the arrow of time is a direct consequence of the expansion of the Universe (Gold 1962). Given the contrast between myriad hot stars and the vast, cold spaces surrounding them, the Universe en masse is well removed from a state of thermodynamic equilibrium. Whereas the stars are obviously sources of energy, the dark realms of interstellar space are perfect sinks—cold reservoirs into which stellar photons irreversibly flow, the radiation apparently absorbed without limit. Universal expansion is a central factor in the three main chapters of this book, all the while attempting to justify mathematically the scenario of cosmic evolution, broadly conceived yet rich in detail.

Other researchers look to various evolutionary records for insight into the meaning of time's arrow (Eldredge 1985). For example, fossils embedded in rocks strongly imply that complex life arose from simple life; the old rocks harbor only fossils of simplified life whereas younger rocks show more complicated remains. Moreover, such evolutionary records are produced not only by biological systems but also by inanimate objects. The stratigraphic ordering of Earth's rocks is widely regarded as a chronological record. The Moon's cratered surface provides a history of its past. The changing internal structures and chemical compositions of stars record the process of their aging. And the varied morphologies among the clustered galaxies suggest evolutionary events that granted them shape. Still, it is the expansion of the Universe that provides a temporal perspective for all inorganic evolution, which ultimately gave rise to all organic processes, including biological and cultural evolution. These now manifest in our conscious awareness of time itself.

What is more, most observed evolutionary trends are irreversible. Although fossil remains of both vertebrates and invertebrates imply that structures or functions once gained can be lost, those lost can seldom be regained. Evolution is a consequence of many variations, chancy to be sure, but occurring in a definite succession, and for it to be reversible, a highly improbable recurrence of specific variations would need to act in an inverse fashion to those that brought about the original transformations.

Complexity itself, however, is insufficient to demonstrate the direction of evolution, at least not biological evolution. Not all species have become increasingly complex; sponges, roaches, spiders, and bees, among numerous invertebrates, are trapped in an endless cycle of perfected daily routines and thus have remained virtually unchanged for eons. Instead, as we shall see, the "direction" of biological evolution probably obeys some rule akin to the following basic maxim: The collective efforts of living organisms tend to utilize optimally (per unit volume or mass) both their energy intake and their use of free energy by dissipative (dispersive) processes occurring within them. It is in this sense that biological evolutionary events have some temporal direction.

Likewise, the release of energy from the Sun and stars suggests a more widespread process unidirectional in time. Solar heat and starlight are created and maintained by the conversion of gravitational into nuclear energy, which in turn emits radiation. Depending on the mass of the star, fusion reactions continue steadily for billions of years, but without any known compensating process capable of refueling cosmic objects, fusion cannot continue indefinitely. Stars, too, are not eternal. And their finite, albeit astronomical, lifetime is enough to impart a temporal history to the Universe. Indeed, the above-stated maxim seems to apply generally to all of cosmic evolution: Organized systems of a physical, biological, or cultural kind everywhere apparently optimize (per unit volume or mass) the flux of energy passing through them. By "optimum" is meant an intermediate range in energy flow, below which the energy flow is too small to affect order and above which is so large as to destroy order—much like either too little or too much food can deprive a person of good health.

Both terrestrial and celestial scales, then, show abundant evidence of a temporal trend in the Universe—provided we consider sufficiently long durations. Macroscopic events in Nature seem everywhere to be irreversible. And so now do microscopic events as well. In recent experiments at the Conseil Européen pour la Recherche Nucléaire (CERN), a huge underground accelerator laboratory spanning the Swiss-French border near Geneva, particle physicists have detected a definite asymmetry in the arrow of time (Angelopoulos et al. 1998). They did so by colliding relativistic protons head-on and then monitoring among the debris the decay of a certain odd particle, called a neutral kaon ($K^0$). In a mere fraction of a second, the kaon changed into its antimatter opposite, an antikaon ($\overline{K}^0$), that is, $K^0 \rightarrow \overline{K}^0$. Also observed in the debris was the opposite decay, $\overline{K}^0 \rightarrow K^0$, which is simply a time-reversed process, much as if the clock were run backward. Enough collisions were created in the accelerator to observe these decay events 1.3 million times, both to build up statistics and to make no mistake about it. The result is that there is nearly a 1 percent greater probability that an antikaon will transform into a kaon than the other way around. It seems that Nature is "naturally" asymmetric in time, even at a most fundamental level, and now we have direct evidence that time's arrow has a preferred direction.

So, while some thinkers—and not just scientists—seek to derive time's arrow from a basic feature of the physical laws themselves, others strive to describe time's asymmetry more in terms of a single basic quantity such as entropy, or a single phenomenon such as cosmic expansion or elementary-particle decay, or even in terms of a process like the flow of free energy through systems that are "open" to their surrounding environments. The French philosopher Henri Bergson (1940), to cite but one example, championed the last of these views, presaging in a few short words the expanded theme of the present work (although he was also positing an *élan vital,* which I reject): "Whenever something is alive, there is open, somewhere, a register in which time is being inscribed." Others demur, denying that time is real; thus spoke the Roman poet Lucretius some two thousand years ago: "time itself does not exist; but from things themselves results a sense of what is to come." Perhaps the essence of time is its transience, an ephemeral quality that will likely prevent its nature from ever being explained in terms of anything more fundamental.

## Quantum Principles

Classical, strictly deterministic physics has now fallen, and with it the mechanical, Newtonian concept that every microscopic event can be precisely described. Whereas Isaac Newton's seventeenth-century view stipulated the physical world as a closed system dominated by cause and effect, we now acknowledge the apparently insurmountable problem that the basic data for *every* particle of matter cannot be completely specified. After all, a mere drop of water contains some $10^{21}$ molecules. Macroscopically, Newton's principles work better (indeed we later use them to explore cosmological models), but even there we run into difficulty: There is no inherent arrow of time in any Newtonian system; everything is reversible, which is contrary to what we see on Earth and in the Universe beyond.

Early in the twentieth century, when scientists began probing the microworld, trying to locate, isolate, and measure the elementary particles of matter, they were surprised to learn that the domain of the atom differs fundamentally from the realm of terrestrial familiarity. Largely through the efforts of the German physicist Werner Heisenberg, the pioneering quantum mechanics of the 1920s reached a consensus that objective observations of Nature's most basic entities are impossible. The very act of atomic and subatomic observation interferes with the process of measurement, significantly altering the state of the object observed. In our attempt to decipher natural order, we have become, as the great Danish atomic physicist Niels Bohr put it, "both spectators and actors in the great drama of existence." We simply cannot separate ourselves from the world around us, regardless of how hard we try.

Whereas classical physics sought to determine precise values for certain physical variables—for instance, the position and velocity of a particle—quantum physics postulates that pairs of these properties can be simultaneously observed only to within a certain limiting accuracy. One or the other can be measured arbitrarily well at any moment, but not both. To perceive much of the quantum microworld, direct collisions of subatomic particles (orchestrated in high-energy accelerators) render intelligible the way particles behave (mostly scatter) among the resulting debris; the analogy often used is that of trying to decipher the interior workings of an intricate watch by smashing two of them together and examining their component parts. In doing so, physicists have invariably found that measurements fluctuate about average values and that these fluctuations arise not so much from practical imperfections of the experimental equipment as from the fuzziness introduced by the act of measurement itself. The resulting premise—the Heisenberg uncertainty principle—to this day inherently constrains the goals of classical physics. Gone is the deterministic and mechanistic paradigm, as well as the notions of strict causality and predestination, that characterized physics for several centuries since Bacon, Descartes, and Newton. Gone, also, are the ideas that the future is implicit in the past and that the world has no essential novelty. No longer can we predict, even if an initial set of conditions are known (let alone chaos science's sensitivity to initial conditions; see Chapter Two, note 4), the one and only final state of some event. In quantum physics, there is no single final state but only several possible alternative states. Probabilities can be assigned to each of the possible outcomes if we are given the initial conditions, but the outcome is not fixed or predetermined. Even Einstein was apparently wrong; "God" *does* seem to play dice with the Universe, thereby creating much originality and diversity.

Of course, averages can be established within the context of large numbers. In this way, statistical statements can often be made with a high probability of success. For example, averages for a stable (equilibrium) system suffice to predict that system's macroscopic behavior with virtual certainty. By contrast, as we shall see toward the end of this Introduction, such certainty does not necessarily hold for unstable systems. Non-equilibrium systems are affected by microscopic disturbances that can accumulate so that events on the macroscopic level reflect chance activity at the microscopic level. Expressed another way, although the uncertainty principle was once thought to be of no importance for the description of macroscopic objects, such as living systems, recent studies of the role of minute fluctuations in non-equilibrium systems imply that this might not be the case; randomness seemingly retains some influence on the macroscopic level as well (Luchinsky and McClintock 1997). Chance behavior of individual particles in the microdomain can surprisingly yield observable changes in the macrodomain.

## Chance and Probability

Chance is surely a factor in all aspects of cosmic evolution, but it cannot be the sole instrument of change. Take, for instance, the issue of galaxy formation. Even as long ago as 1692, Newton reasoned (correctly) that a uniform static cloud of gas will, naturally and of its own accord, randomly develop density inhomogeneities here and there. ("[I]f matter were evenly disposed throughout an infinite space . . . some of it would convene into one mass and some into another, so as to make an infinite number of great masses scattered at great distances from one another throughout all that infinite space"; see, for example, Munitz 1957.) Einstein's theory of relativity (our modern concept of gravity) supports this tendency of a uniform gas to break up, even if expanding. Solely by chance a cloud's matter spawns gravitational fluctuations that become the seeds for the galaxies themselves. Such random fluctuations obey well-understood statistical laws, one of which states that the magnitude of the fluctuation goes like $N^{1/2}$, where $N$ is the average number of gas particles in a system. Thus, if the system in question is a galaxy, wherein the member atoms typically number $10^{68}$, then random fluctuations are likely, at any given place or moment, housing about $10^{34}$ atoms. This makes the fractional fluctuation, $N^{1/2}/N$, equal to a minute $10^{-34}$, which is far too small for detectable inhomogeneities to have grown into galaxies in the time available since the beginning of the Universe. Alternately stated, if the total number of atoms ($\sim 10^{68}$) required to fashion a typical galaxy are to be collected exclusively by random encounters of gas particles, then we run into a problem: A chance accumulation of such a vast quantity of atoms would take several tens of billions of years, implying that few such galaxies should now exist in a Universe roughly 12 billion years old. The observational evidence that galaxies do in fact exist, and richly so in every direction observed, strongly implies that chance could not have been the only factor governing the origin of these magnificent systems. Chance and the gravitational force did each surely play a role, especially in the initial process that triggered the fluctuation and subsequent fragmentation of the cloud, but other agents, such as perhaps turbulence and shocks, must have accelerated the growth of the inhomogeneities so that myriad galaxies could have formed within a time scale shorter than the age of the Universe; in fact, the enhancement process must have been highly efficient since observations strongly imply that all galaxies are old and thus formed long ago.

We need not look into space to find examples of unlikely events that actually did occur. Ample evidence here on Earth exists to show that natural phenomena can constrain chance, producing oddities in structure for which the a priori chances are slim. Just to share a favorite example, most of the known rubies on Earth are found in one hill in Burma—a highly improbable occurrence yet apparently the product of an unexpected series of transformations in mineral chemistry. Life itself may have arisen on Earth (or elsewhere) by means of such an unlikely concentration of chemicals, yet one also unlikely to have occurred by chance alone.

Formation of the precursor molecules of life's origin provides another illustration of the limited role of chance in Nature. In a notable laboratory experiment first performed nearly a half century ago by chemist Stanley Miller (1953), simple molecules such as ammonia, methane, and water vapor interact with one another in an enclosed test tube open just enough to allow energy to enter via a spark discharge. The resulting products are larger molecules, but not just a random assortment of such molecules; among them are many of the 25 amino acids and nucleotide bases common to all life on Earth. More recent experiments show that even with prebiotic gases containing other molecules perhaps more typical of the time (carbon dioxide replacing methane and molecular nitrogen instead of ammonia) and provided that these gases simulating our primordial atmosphere are irradiated with realistic amounts of energy in the absence of free oxygen, the soupy organic matter trapped in this chemical evolutionary experiment always yields the same kinds and proportions of amino-acid-rich, protein-like compounds—and in approximately the same relative abundances found embedding meteoritic rocks and composing living organisms. The specific environment for life's origin is only weakly relevant here, provided it was rich in $H_2$ and therefore reducing (Holland 1984). It is unresolved whether the environment was an alternately violent and then calm primordial atmosphere-ocean as commonly invoked, or a mild and soupy "warm little pond" as Darwin himself envisaged in a tidal basin undergoing gentle and repeated wet-dry cycles, or a more extreme place such as a boiling, sulfurous pool or a hot, mineral-laden, deep-sea volcanic vent, or even the harsh environs of interstellar space that delivered life's building blocks to Earth via comet or meteorite. The point is that if the original reactants were forming and re-forming into larger molecules by chance alone, that is, in the hypothetical case of no forces at all, the products would be a hopeless mess comprising billions of possibilities and would likely vary each time Miller-type experiments were run. But the results show no such chemical diversity, nor does life at the biochemical level. Specificity and reproducibility are hallmarks of this classic experiment, ensuring that bonding of the amino acids does not occur entirely at random, but that certain associations were advantaged while others were excluded. In other words, these prebiotic experiments do not make a little bit of everything; rather, they yield, among other molecules not known to partake of life, significant concentrations of organic compounds that are indeed prominent in living organisms. Selective assembly seems more of a rule than an exception, a proposition to which we shall return in the Discussion.

Of the multitudes of fundamental organic groupings at the DNA and protein levels that could possibly result from the random combinations of carbon, hydrogen, nitrogen, oxygen, phosphorus, and sulphur ("CHNOPS")—estimated and mostly synthesized by pharmaceutical firms worldwide to comprise some 10 million structures containing a hundred carbon atoms or less—fewer than 2,000 are actually employed by most life on Earth (Chothia 1992). These modular parts, or families, which comprise the essence of the approximately 100,000 mosaic genes and proteins of all cellular systems (and despite the millions of known biological species), are in turn based upon only about 60 simple, building-block organic molecules, key members of which are the aforementioned 25 acids and bases (plus another 35 lipids and sugars). More than 90 percent of all living matter is comprised of less than 50 compounds and polymers of those compounds. Some factors in addition to chance are clearly involved in the prebiotic chemistry of life's origin, though one need not resort to supernatural phenomena. Among those factors naturally at work among the molecules are the electromagnetic Coulombic force and non-covalent interactions such as H-bonds, $\pi$-stacking, and van der Waals forces—all of them agents that guide and bond small molecules into the larger clusters appropriate to life as we know it, thus granting the products some specificity and stability. A benzene ring, for instance, is a good deal more stable than a linear array of the same atoms and molecules.

But it's not just microscopic forces at work here; hydrophobic effects and entropy itself often play roles in the construction of small-molecule structural components having an array of sizes, shapes, and polarities. Assemblies of macrostructures (largely via dehydration condensation, that is, the removal of $H_2O$), including the folding of polypeptide chains, are the result of the natural tendencies of water molecules to seek a state of maximum entropy and the chain's component parts to attain a state of maximum (or optimum) free energy. The spontaneous folding and three-dimensional conformation of polypeptide chains (whose details are only poorly known) are less imposed by external forces than by the thermodynamics of open systems that allow morphogenesis in aqueous environments. Apparently each of the chemical evolutionary steps toward life yielded new states more thermodynamically stable than their precursor molecules, accompanied by notable increases in entropy levels of the watery surroundings.

Furthermore, it doesn't take long for reasonably complex molecules to form once the conditions are conducive, not nearly as long as probability theory predicts for a chancy assemblage of atoms. In short, though without detailed understanding (if only because this is a many-body problem), a combination of electrostatics and thermodynamics acts as a molecular sieve or probability selector—in effect, a mark of determinism—fostering certain chemical combinations and rejecting others, all the while guiding organization and order from amid some of the randomness and chaos. The uniformity and ubiquity of life's essential biochemistry, despite the rich diversity of resulting biological types, speaks volumes about the likelihood of an underlying factor, principle, or process—if we could only find it—that drives change.

Such constrained biochemical compounds bring to mind other limitations among Nature's basic features, such as the surprisingly small number of chemical elements. In an argument often attributed to the French mathematician Henri Poincaré, if there were millions (or even only thousands) of elements, rather than the hundred or so known, the resulting combinations based on chance would be so vast as to make it nearly impossible to generalize knowledge about even simple physico-chemical configurations. However, since the number of elements is small, we now know that their various configurations comprise virtually everything we see, feel, or smell. And it is their smallish numbers that have enabled scientists to formulate basic principles of physics and chemistry to describe the order and behavior of real, observed substances, including the laws of conservation, symmetry, and dynamics governing the world and cosmos around us. Likewise, in biology, it's not a question of why there are so many species. Given the genomic wealth and the many possible combinations therein, we might more properly wonder why there are not many more species.

Molecules more complex than life's elementary acids and bases are even less likely to be synthesized by chance alone. For example, the simplest operational protein, insulin, comprises fifty-one amino acids linked in a specific order along a molecular chain. Using probability theory, we can estimate the chances of randomly assembling the correct number and order of acids; given that twenty amino acids are known to partake of life as we know it, the answer is $20^{-51}$, which equals approximately $10^{-66}$. As the inverse of this is obviously a great many permutations, the twenty amino acids could be randomly assembled trillions upon trillions of times for every second in the entire history of the Universe and still not achieve *by chance* the correct composition of this protein. Clearly, to assemble larger proteins and nucleic acids, let alone a human being, would be vastly less probable if it had to be done randomly, starting only with atoms or simple molecules. Not at all an argument favoring creationism, spiritualism, mysticism, and the like, rather it is once again the natural agents of order that tend to tame chance.

Regarding life itself, it is the well-known process of biological natural selection that acts as a sifting mechanism—an "editor" of sorts—permitting some species to thrive even while others perish. Chance admittedly provides the raw materials for biological evolution (especially when environments change so unpredictably), but natural selection is a decidedly deterministic action that directs evolutionary change. The deterministic part can be considered a constraint—a physical, chemical, or environmental boundary—that limits the role of chance; thus we have evolution by "constrained contingency," reminiscent of the Hegelian triad of thesis, antithesis, and synthesis that has infused science for decades and indeed pervades our text. Contrary to popular opinion, and leaning a bit on the coexisting poles of chance and necessity proposed by Democritus of old Greece, Darwin never said that the order so prevalent in our living world arises exclusively from randomness.[1] Ernst Mayr (1997), one of the chief architects of biology's modern synthesis (neo-Darwinism), said it as clearly as can be: "It was Darwin who found a brilliant solution to this old conundrum: [changes in the world] are due to both. In the production of variation chance dominates, while selection itself operates largely by necessity." Yet even the limited role of chance in neo-Darwinism, when coupled with the principle of natural selection, is capable of generating highly improbable, and impressive, results. Theodosius Dobzhansky (et al., 1977), another major contributor to the modern synthesis, also stressed this delicate interplay of chance and necessity (or "trial and success," as he called it), of mutation and selection, of freedom and restraint: "Evolution is a synthesis of determinism and chance, and this synthesis makes it a creative process. Any creative process involves, however, a risk of failure, which in biological evolution means extinction. On the other hand, creativity makes possible striking successes and discoveries."

Crystal growth is a case in point, granting an intuitive feeling for the development of both living and non-living structures via the twin agents of chance and necessity. If the environmental conditions are right, ice crystals will condense out of water vapor, some bigger than others and all a little different; the crystals order at the expense of the surrounding water molecules, whose entropy rises as heat lost to the crystal system enters the environment beyond. One such condition is temperature, of course, and the degree to which it is less than 273 K fixes the geometry of the crystal (flake, granule, rhombohedron, etc.); other key factors include humidity, density, and mass-energy transport (in effect energy flow). To grow such a crystal, water molecules must collide with the crystal in such a way that they stick and are not rejected. The initial molecular collisions are entirely random, but once they occur the migrating molecules are then guided by well-known intermolecular (i.e., electromagnetic) forces into favorable positions on the surface. Only a few symmetric growth patterns are feasible, each pre-established by the deterministic, electromagnetic force. If the incoming molecule lands at a surface position physically conducive to the growth of ice crystal structure, it is "selected" to stay and contribute to the crystal; otherwise it is expelled. Its arrival is random, but the result is not. Nothing designed or purposeful governs the process of crystal growth. Nor is there any reason to suspect that ontogeny and phylogeny among life's diverse forms have any design or purpose, expressed or implied.

A half-century ago, Hermann Weyl (1949) articulated with elegance and beauty the following remarkably prescient, albeit qualitative, statement, as valid today as any regarding both the limited role of chance and the essence of what we today call cosmic evolution:

The statement that the natural laws are at the bottom not only of the more or less permanent structures occurring in nature, but also of all processes of temporal development, must be qualified by the remark that chance factors are never missing in a concrete development. Classical physics considers the initial state as accidental. Thus "common origin" may serve to explain features that do not follow from the laws of nature alone. Statistical thermodynamics combined with quantum physics grants chance a much wider scope but shows at the same time how chance is by no means incompatible with "almost" perfect macroscopic regularity of phenomena. Evolution is not the foundation but the keystone in the edifice of scientific knowledge. Cosmogony deals with the evolution of the universe, geology with that of the Earth and its minerals, paleontology and phylogenetics with the evolution of living organisms.

## Statistical Principles

Given that the element of chance has *some* role to play in all events, we are left with the notion that only when vast numbers are involved do events begin to show expected average patterns consistent with terrestrially familiar behavior of matter in the macrocosm. In this way, we can understand how random processes can yield predictable order, though only when working with large numbers and repeated trials, much as probability theory can elaborately predict the results of tossing numerous pairs of dice. Even so, a price is paid: Whenever a system is represented by an average, some information is inevitably lost about the details of the total system. As a measure of the information lost when working with averages, it is customary to specify the number of individual cases used to determine a particular average.

These statements can be clarified by invoking the nonetheless abstruse subject of *statistical* mechanics, which supplements the classical, mechanistic physics of old. Introduced little more than a century ago by Austrian Ludwig Boltzmann and the first great American theoretical physicist J. Willard Gibbs, statistical mechanics employs large aggregates of particles to represent thermodynamic concepts. Of paramount import, in 1877, Boltzmann proposed his famous entropy formula (a version of which is carved on his gravestone in Vienna),

$$S = k_B \ln W .$$

Here, $k_B$ is known as Boltzmann's constant ($1.38 \times 10^{-16}$ erg $K^{-1}$), ln is the natural logarithm to the base e, and W is the number of different ar-

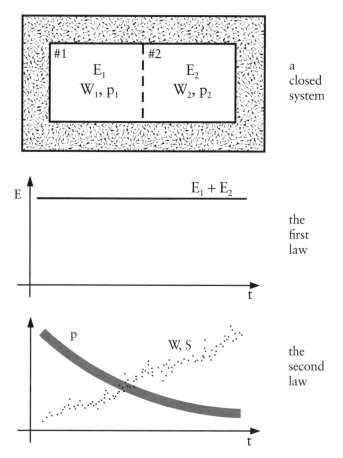

Figure 8. A body of energy, $E_1$, having $W_1$ microscopic states and a probability, $p_1$, of the occurrence of an individual state, in contact with another body characterized by $E_2$, $W_2$, and $p_2$, has a total system energy equal to the sum of the two energies, $E_1 + E_2$, but a total number of microscopic states equal to the product, $W_1 \times W_2$. Thus, the system's entropy, $S = k_B \ln W = -k_B \ln p$, can increase dramatically with microscopic disorder among those states. All this statistical reasoning accords well with the essence of the classical ideas discussed earlier regarding Figure 4; but here, it is *statistical* reasoning, not strict determinism.

rangements of the microscopic states, or "complexions"—positions, velocities, compositions, in addition to various quantum properties—of the individual parts comprising a given macroscopic system.

As an illustrative example, imagine two bodies (1 and 2) in contact, as shown in Figure 8. Assume further that both bodies are immersed in a perfectly insulating container, so they can exchange heat—but only heat, no matter. Now, at any instant, each body will have some energy, say $E_1$ and $E_2$. The first law of thermodynamics demands that the sum of these two energies, $E_1 + E_2$, is constant at all times. Still, at any one time, there will be $W_1$ states of the microscopic entities comprising Body 1 and likewise $W_2$ microscopic states for Body 2; here $W_1$ is a function of $E_1$, and $W_2$ a function of $E_2$. The total number of states for the *whole* system is then $W = W_1 \times W_2$, a multiplicative result since *each* state in Body 1 can be associated with *each* state in Body 2. The multiplicative nature of $W$, in contrast with the additive nature of E (or S), accounts for the logarithmic function in the above Boltzmann formula. It is not unlike the way betting odds are established at the race-track; we know that our chance (probability) of picking a given horse to win *or* place is the sum of the individual odds, whereas our chance of picking two horses to win *and* place is their product.

The role played by statistics can be further appreciated by recognizing that in any macroscopic system the number of microscopic states, $W$, essentially measures the inverse probability, p, of the occurrence of those states. That is, $p = W^{-1} = 1$ for a 100 percent, completely certain state. This is true because the most likely configuration of the particles in a system is the high-entropy state for which they are evenly mixed up or "disordered"; hence the probability is low for any particle to be in a specific microstate. Likewise, during a poker game, the likelihood of certain cards being involved in a desirable hand is high since the entropy of such a preferred or "ordered" set of cards is considered low. Regarding the two-body system described above, the most likely energy distribution is that for which $W$ is maximized, a condition realized when the temperatures of the two bodies are equal. Thus, we can rewrite the second law of thermodynamics as[2]

$$S = k_B \ln(1/p) = -k_B \ln p,$$

and interpret it in the following way: Any isolated system naturally tends toward an equilibrium state of minimum microscopic probability—namely, a uniformity of temperature, pressure, chemical composition, and so on. And since ordered molecular states (for example, where molecules in one part of the system have one property value, but those in the remaining part have another) are less probable than those of random or disordered states, Boltzmann's law then signifies that ordered states tend to degenerate into disordered ones. This law also explains why any material system's available (free) energy tends to diminish with time; useful energy is orderly (or high-grade) energy, whereas heat, associated with the random motions of large numbers of molecules, is usually disorderly energy. Therefore, the energy of an isolated system indeed remains constant but tends to become asymptotically less free to perform useful work as more of it changes into low-grade heat, thereby increasing the disorderliness of its component parts.

This is how the principle of increasing entropy has come to be regarded as a measure of the disorder of a system. The idea of increasing entropy has ceased to be an inviolable law of Nature; rather, it is a statistical law. Even in equilibrium thermodynamics, a reversal of the usual trend—toward a state of lower entropy—is no longer impossible but only highly improbable. For example, a portion of the water in a kettle atop a fire could (theoretically) freeze while the remainder boils, although this is so unlikely as to be considered impossible in our practical experience.

Thus, the static Newtonian view that treats (idealistically) all substances as stable, fixed, isolated components of matter has given way to the conception that all (realistic) phenomena are intrinsically unstable and share part of a dynamic flow. Much like Heraclitus of Greek antiquity, we no longer regard things as "fixed" or "being," or even that they "exist." Instead, everything in the Universe is "flowing," always in the act of "becoming." All entities—living and non-living alike—are permanently changing.

Now, since all things are made of matter and energy, both of which regularly change, we can conclude that the process of becoming is governed by the laws of thermodynamics. As noted in our earlier discussion on time, the second law determines the *direction* in which matter and energy change (or flow), but it cannot determine the rate of that change. In Nature this rate fluctuates, according to the way a system interacts with its environment; even close to home, for this too is part of Nature, earthly events have accelerated in recent times owing to the cultural impact of human beings. There is nothing smooth about the ebb and flow of the becoming process. Change proceeds in jumps and spurts or, to use a fashionable term borrowed from recent studies in biological evolution, in a "punctuated" manner. Accordingly, the process of change is better characterized by non-equilibrium thermodynamics, and we use this relatively new discipline to simulate what *might* happen during any given event, what *might* result from any given change.

## Maxwell's Demon

Thermodynamicists describe order in terms of the arrangements of the numerous microscopic states comprising the various macroscopic properties of a system. The larger the number of possible arrangements, the greater the disorder or randomness of the system. The smaller the number, the greater the order, for order itself can be regarded as a restraint on the way in which a system can arrange itself. As an example, when a system has only a single possible arrangement for its microscopic states, both W and p are unity, and Boltzmann's entropy formula reduces to $S = k_B \ln 1 = 0$, meaning that the system has zero entropy. Admittedly this is a hypothetical case since only the severely impaired behavior of the atoms and molecules in a perfectly ordered crystal at the unattainably lowest possible temperature (i.e., 0 K) could truly have zero entropy—a statement sometimes called the third law of thermodynamics.

The statistical character of the second law arises from the fact, as noted earlier, that the laws of probability govern the direction in which natural processes occur. This is perhaps best understood once we realize that the microscopic properties of individual molecules cannot be dictated by some human agency. The traditional example of "Maxwell's demon" is an appropriate case study. As devised by the nineteenth-century Scottish physicist James Clerk Maxwell, a vessel containing a gas is divided into two compartments by a partition having a trap door controlled by a superhuman demon or robot that can distinguish individual molecules. The proposition is that such a microrobot could open the trap door only when fast-moving molecules approached from one side of the door, thereby allowing the faster molecules eventually to collect in one compartment and the slower ones in the other. Without having expended any work or applied any energy, this exercise would thus raise the temperature of the compartment housing the swifter molecules while lowering the other—namely, cause a change from disorder to order, in apparent violation of the second law.

Doubtless we could squirm out of Maxwell's paradox by affirming the practical impossibility of such demonic robots. But such an assertion would merely dodge the issue, admitting its possibility in principle, even if not in practice. The paradox was solved (Szilard 1929) to the satisfaction of most upon realizing that even such a hypothetical robot would need to know when to open and close the trap door, for knowledge of the velocity of each approaching molecule is required if a segregation is to result between the fast- and slow-moving molecules. In short, the sorting demon would need to have a way to obtain *some information*. For instance, we might imagine that a mini-flashlight (or microscopic radar beacon) could be used to measure each molecule's velocity, thus providing the robot with essential data about whether or not to open the door. This method would work in principle, but it introduces external energy into the vessel. After all, the energy needed to operate the flashlight (or any other apparatus capable of providing the needed information) must come from outside the enclosed container.

Actually, in the contemporary version of the demon—as a computing automaton—it is not the act of information gathering as much as its erasure that dumps entropy into the environment. Before acting the demon-robot would have to store its measurements in its memory, after which the second law would exact its toll when the bits were erased to a perfectly ordered blank sequence of zeros. Hence a minimum energy cost is incurred during erasure (Landauer 1961; Bennett 1982), an irrevocable demand of Nature. Alas, even a technically savvy robot *in an isolated system* could not arrange for a separation of the molecular velocities. Thus, once again we conclude that an energy flow through an open system is an absolute necessity if order is to be created from disorder. And provided we view the process of change sufficiently broadly, there is no contradiction with the fundamental laws of thermodynamics.

Note also that should such a system be open to permit energy to enter from the outside, the demon, now allowed to sort in fact, is actually selecting. A primitive selection process is at work, an apparently necessary condition for the decrease in entropy, indeed for the onset of order. It will not be the last time we encounter the role of selection in the drive toward greater complexity, both within and beyond the realms of physics and biology.

## Information Theory

Our detour into the world of Maxwell's demon was not for naught, for the information—or its opposite, the erasure of such—needed to resolve the paradox points the way to a reconciliation of an apparent conflict between the theorized "destructiveness" of the second law of thermodynamics and the observed "constructiveness" of cosmic evolution.

During the last few decades, the concept of entropy has been occasionally used to illuminate aspects of the subject known as "information theory," although not without controversy. In particular, entropy can be regarded as a lack of information about the internal structure of a system, and it is this lack of information that allows a great variety of possible structural arrangements among the system's microscopic states; the positions, motions, and orientations of the many varied parts of a high-entropic system cannot in practice be specified precisely. Since any one of many microscopic states might be realized at any given time, the lack of information (or ignorance) about a system corresponds to what we have earlier labeled "disorder" (or "uncertainty"). For example, when a system is at equilibrium—the state of greatest diversity on the microscopic scale but of greatest uniformity as seen by the human observer on the macroscopic scale—we have the least possible knowledge of how the various parts of the system are arranged, where each one is, and what they are doing. Close to equilibrium, all such microstates have nearly equal (large) probabilities, p, where $p \leq 1$, which is why the information itself, I, varies inversely as the probability:

$$I \propto \log(1/p) = -K \log p \ .$$

The proportionality constant, K, is a dimensionless unit of information commonly taken as 1 bit (for "binary digit"). Popularly known as the Shannon-Weaver law, this equation was first proposed in 1948 by the engineer Claude Shannon, although he himself never spoke of information per se, only of communication. The concept was further developed by the cyberneticist Norbert Wiener (consult Brillouin 1962), and he, in turn, urged that its central term be called neither information nor communication, testifying to its controversial nature even in its pioneering days some fifty years ago.

The qualitative features of information are conceptually useful in describing ordering in general, yet there are some drawbacks. Foremost among these is its meaning; the word information can connote different ideas in different contexts. In ordinary speech, we use this word as a synonym for news, knowledge, intelligence, and so on. But in the more purified world of cybernetics (or communications engineering), stress is placed on the quantitative *flow* of an intangible attribute, also called "information," from transmitter to receiver. It is in this latter context— a message—that we sense a connection between the principles of thermodynamics and those of information science. To see this, consider how the concept of entropy has been applied to the transmission of information by electronic means, especially regarding the use of telephones, radios, computers, and the like. Such transmissions can be adversely affected on occasion by errors and interference of various kinds, the result often being that the final message is less accurate or conveys less information than the original message. This is true whether the information is transmitted via noisy telephones, static-ridden radios, snowy televisions, faulty CD-ROMs, poorly networked computers, or a host of other means designed to move information from one place to another. Communicating a message inevitably results in a change to a state of greater inaccuracy; the integrity of the message is unavoidably corrupted, some information invariably lost. The same applies to the printed media; when newspapers are torn, letters soiled, or books burned, the informational content of the original message declines.

If we associate the loss of information with a decrease of order, or likewise an increase of disorder, and if we note that the above equation for information resembles that for entropy, we can then relate a gain of information directly to *negative* entropy (Jaynes 1957). This last pair of words plays such a pivotal role in information theory that some agreeable researchers have granted them a single peculiar term, "negentropy," a hybrid that other, unimpressed researchers might equate with "thermodynamic parasite." Actually, this controversial concept was popularized by an Austrian quantum mechanic, Erwin Schrödinger, who raised it in his compilation of seminal lectures, *What is Life?* (1944). The controversy stems not only from the tricky semantics surrounding use of the term negentropy, but also because Schrödinger was a physicist doing innovative biology. (He would not be the last.) Expressed another way, when a system is orderly (that is, low in entropy and rich in structure), more can be known about that system than when it is disorderly and high in entropy. And if entropy measures disorder or the lack of information about a system, then negentropy must be a reasonable assessment of the order, or presence of information. All these latter terms—order, negentropy, information—become approximately synonymous for our purposes. In short, if we gain some information about a system, any previously existing uncertainty about that system is diminished.

These ideas—perceptive and helpful to some, contorted and useless to others—precede even Schrödinger, harking back at least to 1930 and the chemist G. N. Lewis: "Gain in entropy always means loss of information, and nothing more. It is a subjective concept, but we can express it in its least subjective form, as follows. If, on a page, we read the description of a physicochemical system, together with certain data which help to specify the system, the entropy of the system is determined by these specifications. If any of the essential data are erased, the entropy becomes greater; if any essential data are added the entropy becomes less." And Wiener himself, in his classic book, *Cybernetics* (1948), wrote: "Just as the amount of information in a system is a measure of its degree of organization, so the entropy of a system is a measure of its degree of disorganization; and the one is simply the negative of the other."

Thus, the external energy needed to operate the optical or radar apparatus used to secure the information for Maxwell's demon effectively poured negentropy into the vessel, thereby creating order by segregating the swifter gas molecules from their more sluggard companions. In Chapter Three, we shall return to this issue to make use of the strong correlation between the concepts of entropy change in thermodynamics and of data exchange in information science. And then, having sucked from it considerable pedagogical insight, we summarily abandon it in favor of another, more intuitive, indeed more robust yet simpler, concept of energy flow as a measure of complexity among all organized systems, from elementary particles to cultured society.

### Dissipative Structures

The search for an integrated, natural explanation for the origin and destiny of ordered structures has challenged the minds of humans at least since the birth of the scientific method in Renaissance times, and especially thereafter during the Enlightenment, if only sparsely since. Even the philosophy of Platonism, harking back some 27 centuries, has come to be associated, in part, with an underlying order, form, or pattern permeating the nature of all things. Life forms, in particular, are highly complex and intricate systems whose substantial order implies much hierarchical organization—structure gradually or episodically acquired by means of long periods of evolution. Lest life's order be neither created nor sustained, this organization must regulate an ample flow of energy (in the form of radiation, or matter, or both) into and out of living systems—a basic biological (verily, physical) process championed by the German-Canadian systematist Ludwig von

Bertalanffy (1932) and later espoused by Schrödinger (1944). Here is Schrödinger, in one of his more poetic passages, treating life as an open thermodynamic system: "[An organism] feeds upon negative entropy, attracting, as it were, a stream of negative entropy upon itself, to compensate the entropy increase it produces by living and thus to maintain itself on a stationary and fairly low entropy level." As we shall see, similar statements regarding energy consumption and organizational growth are equally valid for the ordered structures that are galaxies, stars, and planets.

That this must be true follows from our previously noted (and tested) idea that the usual course of thermodynamics tends to lead systems toward an equilibrium state of maximum disorder. For any isolated (fully closed) system unable to exchange energy and matter with its surroundings, this tendency is expressed in terms of the now-familiar entropy, which, as also noted earlier, increases until maximized. In short, thermodynamics' second law strongly inhibits ordered structures in isolated systems. Consequently, the apparent contradiction between the observed universal order and the theoretical physical laws cannot be easily resolved in terms of the usual methods of equilibrium thermodynamics or even equilibrium statistical mechanics.

To associate order with low-entropy localized structures such as macromolecules, cells, and even whole planets, stars, and galaxies (which are still relatively "localized"), we need to appeal to nonequilibrium systems. For with departures from equilibrium, like the breaking of symmetry, new things can be created. "C'est la dissymétrie, qui crée le phénomène," said Pierre Curie a century ago. From simple bacteria to complex humans, from aged stars to differentiated planets, ordered systems must emerge, in some cases be maintained, and yet still others be reproduced by means of a continuous exchange of energy with their surrounding environments. Such a sweeping proposition resurrects some of Boltzmann's foresight, also a century ago, while speculating about the Universe being "a sea of generalized uniformity, with here and there some temporary statistical ripples representing individual worlds," yet here we stress an essential mechanism that neither Boltzmann nor Curie had envisioned. This mechanism is the inward and outward flow of matter and energy for systems considered "open." The importance of free energy, available to go to work and potentially defeat entropy, though only locally and temporarily, cannot be overestimated (Morrison 1973).

Called by some researchers "dissipative structures" (Prigogine et al.

1972), all ordered objects—living and non-living—more or less maintain their being by means of a regular flow of available energy from their outsides to their insides. The more complex and intricate the structure, generally the more energy intake (per unit mass) needed for sustenance—a key issue addressed quantitatively in Chapter Three. In the process, these structures can export some of their entropy (or dissipate some of their energy, hence their name) into the external environment with which they interact. Accordingly, order is created and often maintained by routine consumption of substances rich in energy, followed by a discharge of substances low in energy. Either "substance" could be pure energy (i.e., radiation) or matter, or a combination of the two.

How does such structuring occur? How can such ordering emerge from a condition where there was otherwise no such thing? We know well that fluctuations—random deviations from some average, equilibrium value of density, temperature, pressure, etc., also called "instabilities" or "inhomogeneities"—are common phenomena in Nature. They inevitably yet stochastically appear in any system having many degrees of freedom. Normally, as in equilibrium thermodynamics, such instabilities regress in time and disappear; they just come and go by chance, the statistical fluctuations diffusing as quickly as they initially emerged. Even in an isolated system, such internal fluctuations can generate local, microscopic reductions in entropy, but the second law of thermodynamics ensures that they will always balance themselves out. Microscopic temperature fluctuations, for example, are said to be thermally relaxed. Nor can an open system *near equilibrium* evolve spontaneously to new and interesting structures. But should those fluctuations become too great for the open system to damp, the system will then depart far from equilibrium and be forced to reorganize. Such reorganization generates a kind of "dynamic steady state," provided the amplified fluctuations are continuously driven and stabilized by the flow of energy from the surroundings, namely, provided the energy flow rate exceeds the thermal relaxation rate. Global, coherent cycling is often the result, since under these conditions the spontaneous creation of macroscopic structures dissipates energy more rapidly than the ensuing, and damaging, heat can smooth out those structures. Furthermore, since each successive reordering causes more complexity than the preceding one, such systems become even more susceptible to fluctuations. Complexity itself consequently creates the condition for greater instability, which in turn provides an opportunity for greater reordering. The resulting phenomenon—termed "order through fluctuations"—is a distinctly evolutionary one, complete with feedback loops that drive the system farther from equilibrium. And as the energy consumption and resulting complexity accelerate, so does the evolutionary process (Haken 1975; Prigogine 1980; Matsuno 1989). We are now into the realm of true thermo*dynamics,* the older, traditional thermodynamics relegated to the status of "thermostatics."

The simplest system that comes to mind is a well-formed rock (with an atmosphere) orbiting at a fixed distance from a star. Such a system is not in equilibrium with its environment, for energy shining on it heats it above the ambient, cold surrounds of outer space. The surface temperature of the rock in circular orbit is constant because the absorption and emission rates of radiation (both time derivatives) are equal. Thus, we can speak of a steady-state temperature of the rock, but not an equilibrium temperature. Suppose, however, that the rock is in an elliptical orbit, so that its temperature varies cyclically; the result is again a stable, methodical process of energy flowing in and out of the system, although the rock's temperature is now more complex while varying regularly with each orbit. Such cyclic behavior is typical of a dissipative structure that organizes (or reorganizes) itself in a way as to minimize the production of entropy—a disequilibrium process that, not coincidentally, mimics the metabolism of many living organisms whose cycles transport energy. Now suppose further, as would be the case realistically, that the star's luminosity changes over time, much as does our Sun, which is brightening at the rate of ~1 percent every hundred million years. The result would be a rock temperature that cycles dissipatively *and also* changes slightly but steadily over long periods of time—a modulated evolutionary change displaying steady drifts in system properties and making it impossible for any cycle ever to repeat perfectly. Non-equilibrium thermodynamics stipulate that radically different states can succeed one another with sudden abruptness, an adaptation to wholly new complex states even as these boundary conditions slowly change. The result, rare and catastrophic, might be violent climate change or major alteration in surface cycling of matter and energy—a largely deterministic response to a decidedly stochastic event, and all explainable (if not predictable) in terms of unchanging scientific principles guiding gradients, flows, drifts, and cycles.

A common solution to the dynamical-system equations governing non-equilibrium states (beyond the scope of this book) is that of a "bifurcation" (Poincaré 1914; Turing 1952). The fork diagram of Figure 9(a) illustrates such a twofold solution, showing how a stable steady state can spontaneously break its symmetry upon entering a non-linear mode beyond some energy threshold. The result is two new, possible, dynamical steady states, only one of which a given system actually selects, depending partly on the system's history and partly on environmental fluctuations at the time of bifurcation. In non-equilibrium thermodynamics, the new states can result in whole new spatial structures at each bifurcation, much as in equilibrium thermodynamics when changing temperatures cause phase changes in various states of matter. As before, it is the (free) energy entering the system from the outside that helps create the conditions and engenders the changes needed to enhance the potential growth of complexity with time. The interplay between random factors (at the bifurcation point) and deterministic factors (between bifurcations)—chance and necessity, again—not only guides the systems from their old states into new configurations, but also specifies which new configurations are realized. However, with statistical fluctuations triggering such events and the role of chance once more evident, the evolutionary trajectories are indeterminate, the new complex states unpredictable. Again, with evolution, we are not in the prediction business; cosmic evolution is an attempt to explain the rich ensemble of past events, whatever their convoluted, historical paths might have been.

Figure 9(a), nonetheless, forecasts an integral part of this book. The environmental fluctuations are the physical equivalent of system mutations, the stabilizing paths taken akin to natural selection. To be sure, transitions to new, more complex states in biological evolution are often represented by stepwise diagrams like that of Figure 9(b), which is adapted from Salk (1982). Here, organisms have evolved traits needed to survive successfully in various niches, labeled A. Nothing much happens ("stasis") until the environment alters ("punctuates"), such as by climate change, competing species, or new predators, at which time a challenge is posed to the continued survival of that species (arrows). Should the species be unable to develop (or evolve) new characteristics allowing it to meet the challenge (new food sources, ability to hide or escape from predators), it will die, C. If, by contrast, its accumulated physical and behavioral traits are conducive to those needed in the changing environment, B, then it will survive, having adapted to its new circumstances, A—at least until some additional threat to its survival appears (arrows), forcing that species to evolve yet again. Much as for physical systems, biological systems do change in response to an unpredictable mixture of randomness (the environmental alteration) and determinism (the species adaptation).

Beyond certain instability thresholds, systems can dramatically

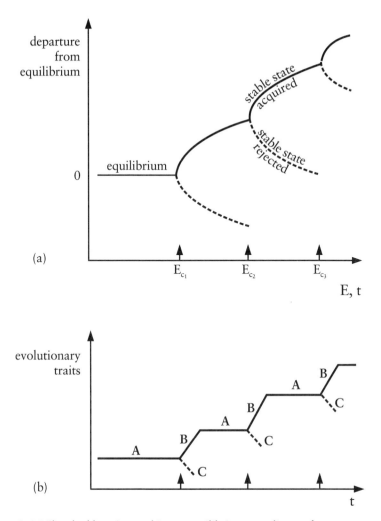

Figure 9. (a) Sketched here is an arbitrary equilibrium coordinate of an open system as a function of both time and energy, either of which serves diagrammatically to illustrate the extent of departure of that system from equilibrium. The time axis makes clear that this is an historical, evolutionary process, whereas the parallel energy axis denotes the free energy flowing through the open system as a vital part of that process. At a certain critical energy, labeled here $E_{c_1}$, a system can spontaneously change, or bifurcate, into new, non-equilibrium, dynamic steady states. Statistical fluctuations affect which fork the system selects upon bifurcation (arrows), namely which spatial structure is achieved, therefore the end result is unpredictable. At right, a second and third bifurcation occur further on in time, with the application of additional energy, $E_{c_2}$, and then $E_{c_3}$. (b) Events in evolutionary biology mimic those of the diagram in (a), here sketched against an arbitrary evolution coordinate. In phases marked A, the main task of a species is to survive and thus persist until such time as the environment changes (arrows), afterwhich further evolution occurs—along phase B toward survival or phase C toward extinction. The upward rising graph implies no progress, but it does indicate a *general trend* toward increasing complexity.

change, fostering the spontaneous creation of an entire hierarchy of new structures displaying surprising amounts of coherent behavior. Such highly ordered, dissipative structures can be maintained only through a sufficient exchange of energy (again, as radiation, matter, and often both) with their surroundings; only this incoming-outgoing energy flow can support the requisite organization for an ordered system's existence. The work needed to sustain the system far from equilibrium is the source of the order. An apt analogy is that of urban organization: A city can survive only as long as food, fuel, and other vital commodities flow in, while products and wastes flow out. Numerous other examples will soon be presented for solids, liquids, and gases, ranging from fluid dynamics to weather phenomena, followed by biological and cultural examples as well. Even mental urges and instincts, ranging from ordinary thoughts to exotic dreams, are regarded by Freudian doctrine as gusts of energy swirling through the brain; Sigmund Freud based his analysis of psychodynamics on hydrodynamic and thermodynamic models (though he wrongly equated good health with equilibrium), embracing in a Heraclitean spirit the importance of "flow," as did William James who may have coined the phrase, "stream of consciousness" (Sabelli et al. 1997). Enhanced energy input to the brain is probably why our extremities often grow cold when thinking quietly or reading intensely; to be sure, Nuclear Magnetic Resonance (or MRI) scans of human skulls show that heightened mental states (especially electrical activity in the visual cortex) require increased blood flow through the brain, which gains its energy from the oxidation of glucose in the bloodstream.

For open systems, then, the second law of thermodynamics must be reformulated. Specifically, we now recognize two contributions to the total entropy change of any event in Nature: One is the (normally positive) entropy production inside the system per se caused by irreversible events, such as friction, chemical reactions, or conduction of heat; the other is the (always positive) entropy production in the environment caused by exchanges between the system and the outside world. Symbolically, the total entropy change,

$$\delta S = \delta S_{sys} + \delta S_{env} \geq 0,$$

where $\delta S_{sys}$ and $\delta S_{env}$ are the entropy changes of the system itself and of its surrounding environment, respectively. Although no formal proof of this inequality exists (or is perhaps even possible), the absence of any experimental exceptions provides confidence in its widescale validity.

For a totally isolated system in which neither matter nor energy is exchanged with the external environment, $\delta S_{env} = 0$ and $\delta S > 0$ (save for the case of an ideal, reversible, and unrealistic process for which it is zero); the total entropy change is then uniquely determined by $\delta S_{sys}$, which necessarily grows as any realistic system performs work. By contrast, for an open system, the entropic change caused by environmental interaction is always positive ($\delta S_{env} > 0$), whereas the internal entropy change ($\delta S_{sys}$) can be either positive or negative (see Figure 7). The sign of the internal contribution to the total entropy change is generally undetermined and depends on the quality and rate of the flow of energy across an open system's boundary. Provided that $\delta S_{env}$ increases more than $\delta S_{sys}$ decreases, the second law of thermodynamics is still obeyed in that the total entropy change $\delta S > 0$, meaning that a system can become locally ordered at the expense of a global increase in entropy in the Universe beyond.

Accordingly, in principle at least, an open system can establish and sustain a negative entropy (or negentropy) change in relation to its environment, and during the course of evolution may reach a state wherein its entropy has physically decreased to a smaller value than at the start. Note that we are not stating that the rate of entropy growth merely becomes lessened or even reduced to zero. Under appropriate conditions, most notably a high-grade energy flow, the entropy itself within open systems can actually be reduced, i.e., $\delta S_{sys} < 0$. For such a non-equilibrium state, order can be achieved within a system by means of the spontaneous emergence of organization.

Furthermore, open systems can be prevented from reaching equilibrium by the regular introduction of fresh reactants and by the regular removal of the derived products. And some such systems can be regulated to achieve a constant ratio of incoming reactants to outgoing products. The system then *appears* to be in equilibrium because this ratio does not change over time. But in contrast with true equilibrium (where entropy is maximized), this kind of process continually produces entropy and dumps (or dissipates) it into its surrounding environment. In this way, both order and entropy can actually increase together—the former locally and the latter globally (Landsberg 1984).

The phenomenon of refrigeration includes a simple example of the creation and maintenance of non-equilibrium order within an open system at the expense of increasing disorder outside that system. Imag-

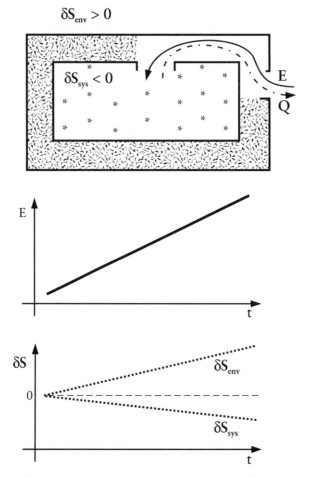

$$\delta S_{env} > 0$$

$$\delta S_{sys} < 0$$

Figure 10. A refrigerated solution of sugar and water exemplifies how energy, E, entering the solution (say, from an electrical outlet) can drive parts of the open system far from equilibrium, thus forming islands of ordered structures known as sugar crystals. The resulting expulsion of heat, Q, to the surrounding environment can cause a lowering of the system's entropy, $\delta S_{sys}$, all the while the entropy beyond the open system, $\delta S_{env}$, is increased; in fact, $\delta S_{env}$ increases more than $\delta S_{sys}$ decreases, keeping the total $\delta S > 0$, in accord with the second law of thermodynamics.

ine, as illustrated in Figure 10, a container of water saturated with sugar. Normally, such a solution comprises a relatively random state, the water and sugar molecules freely able to move about, thus occupying a great many positions relative to one another. As the system cools, sugar crystals begin forming spontaneously, in the process becoming highly organized as individual molecules occupy rather exact positions in the emerging matrix comprising the crystal. Now possessing order, the newly formed crystals themselves necessarily possess lower entropy than the surrounding solution; the act of crystal formation actually decreases the entropy in certain localized parts of the system. To compensate, we should expect to find an increase in entropy somewhere else, but the solution itself is unlikely to have an appreciably increased entropy since its temperature has also been lowered. We need to enlarge our view beyond the confines of the vessel housing the solution, for as the solution cools, heat flows to the surroundings beyond the container. Therefore, it is the air outside the system that must suffer an increase in entropy—which is exactly why an open refrigerator cannot be used to "air-condition" a kitchen in midsummer; a refrigerator, while cooling its contents, in fact tends to warm the kitchen.

Although the *destruction* of order always prevails in a system in or near thermodynamic equilibrium, the *construction* of order may occur in a system far from equilibrium. Whereas classical thermodynamics deals with the first type of physical behavior, novel aspects of non-equilibrium thermodynamics only now being developed are needed to decipher the second type of behavior.

Heating of a thin layer of fluid from below provides a good illustration of such dual behavior (Chandrasekhar 1961), a case technically termed a "hydrodynamics instability" or simply a Bénard cell, after the French physicist, Henri Bénard, who conducted a series of ingenious experiments in thermal transfer in 1900. As shown in Figure 11, externally applied energy generates a vertical thermal gradient in the fluid capable of enhancing any random molecular fluctuations; small pockets of molecular aggregates spontaneously form and disperse. When the heating is slight (i.e., below some critical temperature or instability threshold), the energy of the system is distributed by conduction among the thermal motions of the fluid's molecules and the fluid continues to appear homogeneous; the natural, random fluctuations are successfully damped by viscosity, which prevents pockets of warm water from rising more rapidly than the time required for them to attain the same temperature as their neighbors, and the state of the system remains stable and incoherent, or effectively equilibrated. But beyond this threshold (i.e., when the fluid is heated extensively), instabilities naturally amplify as large thermal gradients develop, thereby spontaneously breaking the initial symmetry (or homogeneity) of the system; this causes the onset of macroscopic inhomogeneities, or bulk mass movements, in the fluid—namely, small but distinctly and coherently orga-

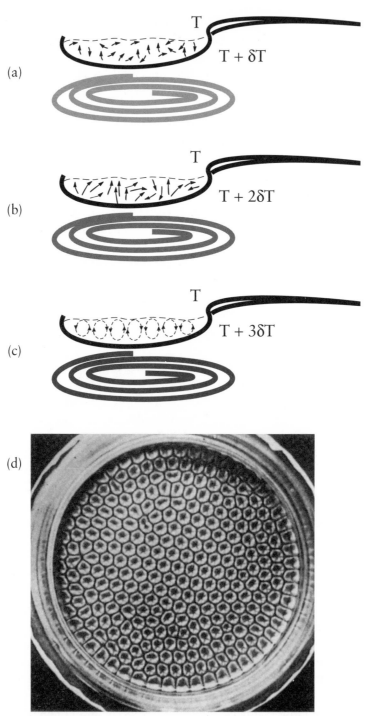

nized eddies that upwell and descend via convection, not conduction. Anyone can verify this well-known convective phenomenon by slowly heating a pot of shallow water on the stove (deep layers of water obscure the effect); as the water is brought to near boiling, cells housing millions of $H_2O$ molecules become buoyant enough to overcome viscosity and to move systematically, thereby forming a stable, circulating pattern of characteristic size. In this way, order can naturally emerge (that is, spontaneously organize) when the system is driven far beyond its equilibrium state. That is what is meant by the principle of "order through fluctuations," or "energy flow ordering." The phenomenon is also sometimes called "self-organization," although that term and others like it (those with the prefix "self-") are deceptive in that such ordering is actually occurring not by itself, as though by magic, but only with the introduction of energy.

How, specifically, do the coherent structures come about, and what is the origin of the sudden disorder-to-order transition? Here is a way of physically intuiting it (Schneider and Kay 1995). The energy flowing into the system first creates a temperature gradient from top to bottom in the fluid. Eventually, as the temperature difference exceeds some threshold (depending on the nature of the fluid and the environmental circumstances), a structure emerges, in this case the hexagonally

---

Figure 11. The emergence of ordered convection (Bénard) cells is illustrated here as a shallow pan of water is progressively heated from below. The temperature T of the fluid layer at the bottom exceeds that at the top by an amount $\delta T$; actually, $\delta T$ determines the temperature *gradient*, technically related in turbulence theory to the so-called Reynolds number (Feynman et al. 1964). In frame (a), random instabilities arise upon slight heating (small $\delta T$), but the water molecules move around aimlessly without any order or pattern; heat flows evenly through the system from molecule to molecule and the system is close to equilibrium. In (b), with greater heating (moderate $\delta T$), the molecules have higher velocities but their motions are still mostly random; the liquid remains largely disorganized and approximately equilibrated all the while beginning to develop increased thermal and density gradients. In (c), as the temperature gradient in the water exceeds some critical value (large $\delta T$), the fluid's molecules depart far from equilibrium and partake of an organized convection pattern; the temperature inversion is now sufficient to set the medium into a series of rolls or hexagons, with the warmer, low-density fluid rising and the cooler, high-density fluid falling in an orderly fashion that is stable in time. Part (d) shows an actual photograph, looking down onto the field of hexagons comprising the convection cells of such an ordered system; once the heat source is removed, the cooperatively ordered movements soon cease.

shaped Bénard cells that enhance the movement of heat. Such structures actually attempt to break down, or neutralize, the gradient by increasing the rate of heat transfer. The newly formed structures themselves, then, are Nature's way of trying to return the system to equilibrium. Provided energy continues to enter the system from outside, the individual structures remain more or less intact. In a crude sort of way, they are selected to endure. In even cruder terms, the regular energy flow resembles a primitive metabolism, somewhat akin to a more highly organized version operating in life forms and examined later in this Introduction (see Case of Cell Metabolism). For Bénard cells, the greater the flow of energy (per unit volume or mass), the steeper the temperature gradient, and the more complex the resulting structure—at least up to some limit beyond which the energy becomes so great that the structure, and perhaps the entire system, are destroyed.

Likewise, coherent, whirling eddies can be generated by slowly passing our hand through a tub of water, or a teaspoon through coffee. Rapid and irregular passage only produces splashing (a kind of turbulent chaos), but a moderate, steady movement yields organized whirlpools, or vortices, of swirling water in its wake. Here the tub of water may be considered an open system, with our hand providing some energy from outside. Without this (free) energy the water would remain idle and quiescent, the epitome of a closed system in perfect equilibrium. But the application of external energy enables the liquid system to generate a density gradient, depart from equilibrium, enhance statistical fluctuations, and, so long as energy is provided, establish somewhat ordered structures such as that in Figure 12(a). By contrast, too much energy would cause the water to spill out or boil away, the medium itself to disperse.

Planetary weather is another practical example of a semblance of order within a complex and chaotic system of gas. Earth's atmosphere is heated largely from the Sun-warmed crustal surface, thus creating thermal updrafts, surface winds, and other meteorological phenomena in a reasonably systematic way. The physical process is, once more, a Bénard instability that determines the motions of air currents and the formation of clouds. Wind itself is an excellent illustration of a semi-organized state, in this case the partly coherent movement of air molecules. Calm air, in contrast, is characterized by random molecular motion of a highly probable, high-entropic gas. Tornadoes are paragons of order-through-fluctuations. Small, naturally occurring wind shear effects, under conditions of severe non-equilibrium and strong pressure gradients, can amplify into massive energy flows that, though superbly (and locally) constructed, can be utterly (and globally) destructive, ravaging the environment to feed the sustaining storm with ever more energy.

An especially apt case of energy-triggered generation of order is provided by the occasional yet sudden interruption of surface winds blowing across our planet's oceans. For example, in the Canary Islands, as shown in Figure 12(b), the prevailing westerlies flowing across the Atlantic Ocean encounter the mountainous regions near Tenerife, causing swirling updrafts and turbulent eddies of moist air to form high in the atmosphere. Such kilometer-sized vortices come and go at random, like bubbles rising in a pot of hot clam chowder; they can be seen clearly by looking downward from orbit, though not easily by looking upward from the ground (owing to low-ceiling clutter). Solar-driven terrestrial winds provide the external energy which, along with the fluctuations in the flow caused by the Canaries, eventually results in a growth of atmospheric structure. Patterns of cumulus clouds develop as rising currents form small competing cumuli that draw on the energy contained mainly as latent heat stored in water vapor. A process of inanimate selection sets in, with those cumuli able to attract more air flow, and thus more energy, causing the demise of others. By the end of a typical hot, sunny summer day, selection has fostered the development of a few large-scale thunderstorms, each well removed from the equilibrium characterizing more normal meteorological conditions. The rising columns are parts of convective cycles whose air also loops back to the ground, thereby comprising a mechanism that reinforces the cloud over the rising column, helping it move higher and increase in volume. Very infrequently, should they be driven far enough from equilibrium and fueled with adequate moisture, some eddies can mature into full-scale hurricanes hundreds of kilometers across like that in Figure 12(c); they have been naturally selected by a well-understood physical process that mixes a random initiator with a deterministic outcome, a theme integral to the onset of order. The thermodynamics of a hurricane can be modeled as an idealized heat engine, operating between a warm heat reservoir at ocean level ($\sim 300$ K) and a cold sink some 15 to 20 km up in the troposphere ($\sim 200$ K). The energy realized, hence the storm's intensity—its strongest wind speed—is proportional to this temperature difference (Emanuel 1999). What's more, the maintenance and probably the origin of hurricanes are also known to occur during seasons of maximal thermal disparity between the tropical and temperate

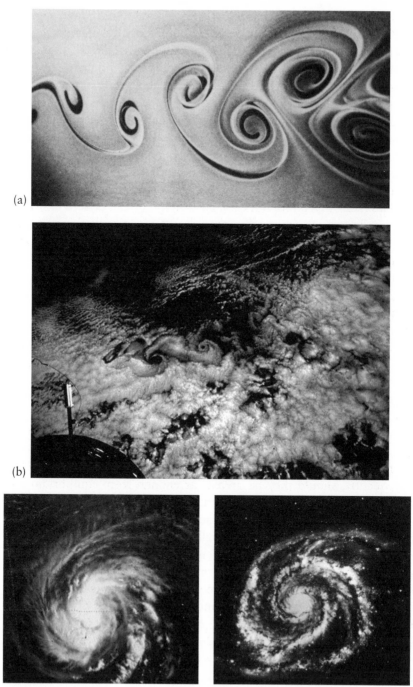

(a)

(b)

(c)

(d)

oceans—in short, to depend on strong gradients (here in temperature and pressure). Even the deadliest hurricanes die out as they move away from the warm waters that power them.

Interestingly enough, the pancake shape, the spiral-arm structure, the distribution of energy, the differential rotation pattern, and many other morphological characteristics of hurricanes bear an uncanny resemblance to those of spiral galaxies such as the one shown in Figure 12(d); even the "eye" in a hurricane conjures up the purported "hole" (black or otherwise) in the cores of most galaxies. Is it possible that the origins of Nature's grandest structures were triggered in the early Universe as the rapid flow of radiation, launched by the big bang, swept past primordial gas fluctuations that acted as the turbulent seeds of galaxies? And since most meteorologists agree that some sort of turbulent "priming" is needed to initiate a hurricane, might not studies of the formative stages of such storms—ironically, in the very air we breathe— conceivably be used by astronomers to derive some clues to the elusive density inhomogeneities that presumably gave rise to protogalaxies some 10 billion years ago?

Other planets, and even the Sun, often display the emergence of order in the form of eddies, vortices, whirlpools, and the like. Saturn, for instance, shown in Figure 13(a), in 1990 spawned a huge white "spot" in its banded but otherwise featureless upper atmosphere. Saturn, like Jupiter, is known to produce internally more heat than it receives from the Sun, so the spot was presumably driven by upwelling convection caused by reservoirs of stored energy beneath the cloud deck. It grew in

Figure 12. (a) Temporarily organized vortices naturally occur in the wake of a rock in a stream, a moving canoe paddle, or a human hand passing through a tub of water. (b) This *Gemini* spacecraft photograph of the *top* of Earth's cloud layers above the Canary Islands clearly shows the presence of several kilometer-sized eddies that come and go in the form of atmospheric updrafts as westerly winds blow across the islands (courtesy NASA). (c) A full-scale hurricane is a huge collection of moisture hundreds or thousands of times larger than an atmospheric eddy, a case of a small pattern having become vastly enhanced owing to plentiful incoming resources and optimal environmental conditions. In this 1976 photograph, a weather satellite captured Hurricane Emmy in the north central Atlantic Ocean (courtesy NASA). (d) A spiral galaxy is a gargantuan collection of stars, gas, and dust spanning 100,000 light-years, or more than a trillion trillion times larger than a typical hurricane; this one, called M51, the Whirlpool Galaxy, is magnificently structured even if not very complex (courtesy National Optical Astronomy Observatory).

a matter of weeks to several times the size of Earth, yet did not last long. Torn by differential rotation patterns but still displaying extraordinarily rich structure never before seen on Saturn, the spot in just a few months broke up into a band-like streak almost completely girdling the planet's equatorial regions. This is probably how the latitudinal bands (which are a kind of structure) arose on each of the Jovian planets, especially Jupiter. Even the Sun, shown highly magnified in Figure 13(b), displays some mildly regular patterning as evidenced by its granulated photosphere. Such surface granulation is only the topmost tier of an entire hierarchy of convection cells stacked atop one another throughout the upper layers of the solar interior. The Sun is heated at its core originally by the gravitational potential energy of its infalling parent cloud, and ultimately its ordered insides arise at the expense of a disordered environment beyond.

The laser is yet another dramatic example of the abrupt emergence of order when an open system is driven far from equilibrium. "Laser" is an acronym for "light amplification by stimulated emission of radiation," a device (human-made in terrestrial applications or naturally operating in interstellar space) able to produce high-quality, coherent light from a cluster of energized atoms and molecules—a virtual plasma. The light emanating from a laser depends critically on the manner in which the incoming energy has excited the gas, as well as on the ambient environment engulfing the gas. Below some instability threshold (corresponding to a small use of energy as from an ordinary AC outlet), only incoherent light results, much like that from any household light bulb. In a hot, nearly equilibrated gas of an ordinary lamp, each of the atoms within the gas emits light randomly and independently. Beyond that threshold (corresponding to a gradual increase above some critical value of the applied or "pumped-in" energy, for example that supplied by a more powerful generator), the plasma within the laser switches spontaneously to the emission of coherent light. Here, an excessive number of atoms leap to an excited state, the whole system becomes decidedly non-equilibrated, and the result is a coordinated behavior of the atoms that emit light precisely in phase. The gaseous system is said to have changed—experienced a phase transition—from disorder to order, in effect punctuating its equilibrium to achieve a new level of organization.

Regarding these and other open systems that are optimally energized in Nature, the Belgian physical chemist Ilya Prigogine (1979) has summed it up smartly: "The atoms and molecules in such systems must interact with their immediate neighbors through well-known, short-

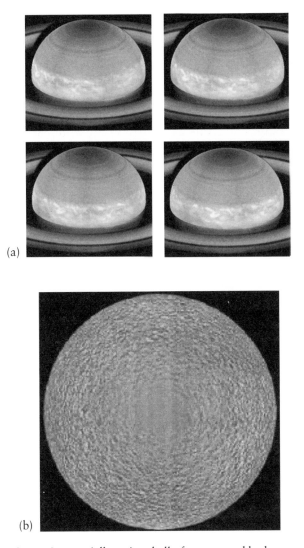

Figure 13. (a) Saturn is essentially a giant ball of compressed hydrogen and helium gases (shown here up close without its famous rings), with just enough methane, ammonia, and other trace compounds to give it some bland color. The planet's white spot, shown here in hourly spaced, time-lapse images (left to right, top to bottom) recorded by the Earth-orbiting *Hubble Space Telescope,* displays much structure even while dispersing around the equatorial regions; the spot's color is apparently due to high cirrus clouds of frozen ammonia crystals. Seen on Saturn at intervals of about 60 years since the mid-nineteenth century, the spot must represent an intermittently massive updraft, like the bubbles in a pot of oatmeal. (b) A filtered photograph of the Sun's granulated photosphere, taken by the space-based *Solar and Heliospheric Observatory,* shows a rich array of convection cells, each comparable in size to Earth's continents (~1000 km across). The bright portions of the image are solar cells where hot matter upwells from below; the dark regions correspond to cooler gas sinking back down into the interior (courtesy NASA).

range chemical and physical forces. Each atom or molecule knows only its immediate neighbors and its direct environment. That much is normal. But in these new [dissipative] structures, the interesting thing is that the atoms and molecules also exhibit a coherent behavior that goes beyond the requirements of their local situations ... which increases in complexity and grows to be something vastly different from the mere sum of its parts."

In the remaining sections of this Introduction, we examine a few representative cases of complex, organized matter. For each case, the extent of entropy decrease in the system itself is calculated, followed by the entropy increase in the environment beyond, thus ensuring that the net change accords well with the second law of thermodynamics.

### The Case of Water Formation

Consider the synthesis of ordinary liquid water at ambient room conditions—298 K temperature and 1 atmosphere (bar) pressure. The basic chemistry involves a spontaneous, exothermal reaction between two simple and obvious molecular gases, the result being the production of a slightly more complex water molecule and some energy:

$$H_2 + \tfrac{1}{2}O_2 \rightarrow H_2O + 2.8 \times 10^{12} \ \text{erg mole}^{-1}.$$

The entropy change within the system ($\delta S_{sys}$) comprising the reaction mixture equals the difference of the initial ($S_i$) and final ($S_f$) entropies of the reactants and product, each of which can be found tabulated as standard molar entropies in many physical-science handbooks (see, for example, Hodgman, et al., any edition):

$$\begin{aligned}
\delta S_{sys} &= S_f - S_i \\
&= S_{H_2O} - (S_{H_2} + \tfrac{1}{2}S_{O_2}) \\
&= -1.6 \times 10^9 \ \text{erg K}^{-1}\text{mole}^{-1}.
\end{aligned}$$

During this six-dimensional (three-space, three-velocity) phase change, the liquid water molecules have had their spatial order increased compared to their prior constituent gas molecules. The system's entropy has decreased since it is clearly more organized than before (indeed, 1.5 molecules have transformed into one).

By contrast, the entropy of the surroundings increases, which is hardly surprising as the thermal energy produced gets dumped there. The released heat broadens the distribution of velocities in that part of the six-dimensional phase space of the remaining gas molecules beyond the system:

$$\delta S_{env} = \delta Q/T$$
$$= 2.8 \times 10^{12} \text{ erg mole}^{-1}/298 \text{ K}$$
$$= +9.5 \times 10^{9} \text{ erg K}^{-1} \text{mole}^{-1}.$$

Hence, the net entropy change in the Universe, for this singular reaction whenever and wherever it may occur, equals the sum of these two changes, or $+7.9 \times 10^{9}$ erg K$^{-1}$ mole$^{-1}$. This is indeed consistent with the second law since the increase in spatial order of the product water molecules has been more than offset by the increase in the spread of reactant gas molecules in velocity space.

Realistically, not all of the 2.8 trillion ergs liberated in the act of water formation (actually a consequence of the first law of thermodynamics) need be delivered wastefully to the surrounding environment. That is because, to compensate for the decrease in entropy of the reaction mixture, $S_{env}$ need not increase by the full $9.5 \times 10^{9}$ erg K$^{-1}$ mole$^{-1}$; an increase of $1.6 \times 10^{9}$ erg K$^{-1}$ mole$^{-1}$ would suffice, and still not violate the second law. Thus, the excess energy produced ($2.3 \times 10^{12}$ erg) could be used in some other way, perhaps to lift a mechanical weight or merely to be stored in a battery. The most common use of the controlled production of water in this way may well be to power an electrochemical cell. A fuel cell whose byproduct is energy and drinking water is very useful on board a crewed spacecraft. It can be shown (Bent 1965) that the maximum voltage produced by such a cell is 1.2 volts. Conversely, we could say that 1.2 volts is the minimum voltage needed to electrolyze (or decompose) a mole of water at room temperature,

$$H_2O + 1.2 \text{ volts} \rightarrow H_2 + \tfrac{1}{2}O_2 .$$

This reaction again proceeds, in this case endothermically, in good agreement with the laws of thermodynamics.

A related calculation further shows that water, once formed in the liquid state, can become additionally ordered when transformed into the solid (ice) state. Briefly, if liquid water at 273 K (0°C) and 1 bar pressure is subject to environmental conditions with T < 273 K (say, 271 K), then ice forms and $6 \times 10^{10}$ ergs mole$^{-1}$ of heat is released to the surroundings. The respective changes in entropy of the system and its environment are:

$$\delta S_{sys} = \delta Q_{H_2O} / T$$
$$= -6 \times 10^{10} \text{ erg mole}^{-1} / 273 \text{ K}$$
$$= -2.20 \times 10^8 \text{ erg K}^{-1} \text{ mole}^{-1},$$
$$\delta S_{env} = \delta Q_{env} / T$$
$$= +6 \times 10^{10} \text{ erg mole}^{-1} / 271 \text{ K}$$
$$= +2.22 \times 10^8 \text{ erg K}^{-1} \text{ mole}^{-1}.$$

The net production of entropy then equals $+2 \times 10^6$ erg K$^{-1}$ for every mole of water frozen, an admittedly slight value yet one in accord with thermodynamic theory and experimental measurement. What it means, frankly, is that each time an ordered ice cube is formed in a freezer (by pumping in energy to keep the machine running), the Universe becomes just a little bit more disordered.

### The Case of an Emerging Star

Until now we have spoken mostly of thermal and chemical (or configurational) entropy. But there is an entropy due to gravity as well, especially for those massive systems dominated by gravitational interactions. And this quantity, at first notice, behaves a little oddly, for gravity displays counter-thermodynamic tendencies. As gravitating systems lose energy they grow *hotter* owing to core contraction. Figure 14 provides some perspective, part of it adapted from the work of British physicist Roger Penrose (1989), who has written a nice description of gravitational entropy (see also Saslaw and Hamilton 1984).

Gravitational entropy is particularly important in understanding, while later discussing the early Universe, how it is that the primordial cosmic fireball began in a low-entropy state, despite its then-prevailing thermal and chemical equilibrium. Suffice it to say now that, for an ordinary gas in a box wherein gravitational effects are negligible, entropy increases impel the gas toward greater uniformity, namely less structure and more disorder, just as we've seen earlier. But for a system of gravi-

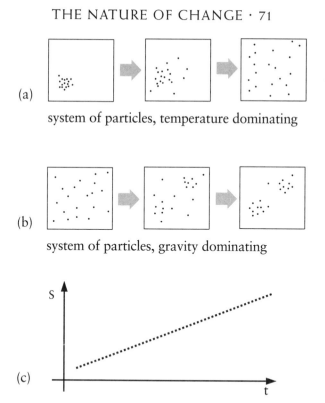

(a)

system of particles, temperature dominating

(b)

system of particles, gravity dominating

(c)

Figure 14. (a) Entropy increases as an ordinary gas disperses in a box having dimensions small enough for gravitational interactions to be neglected. (b) Among gravitating bodies on larger scales, entropy increases when the opposite occurs, namely, when diffuse gas clumps to form ordered structures. (c) For either case, the net entropy rises, but the details and resulting phenomena depend on scale.

tating bodies, the reverse is true, largely because of the unshielded, long-range, and attractive nature of the gravitational force. In other words, a large ensemble of particles in the absence of gravity will tend to disperse, yet in the presence of gravity will tend to clump; either way, the net entropy increases. As implied by Figure 14, entropy gains can be considerable when a massive cloud fragments and contracts into a planet, star, galaxy, or larger structure. Actually, the entropy increase occurs outside such ordered systems; the systems themselves lower their entropy by radiating it away into the surrounding space.

The case of star formation will serve as a numerical example. Like many ordered systems and processes among myriad complex structures in Nature, newly formed stars can be shown quantitatively to have lowered their entropy at the expense of their surrounding environment. Stars, too, are open systems. Here we employ some of the same thermodynamic principles that chemists use to analyze gas dynamics or that engineers use for heat engines.

Consider, initially, a huge interstellar cloud of perhaps light-year dimensions, for all practical purposes structurally, chemically and thermally homogeneous throughout. Such a system is virtually in steady-state equilibrium, its temperature ($\sim$15 K) nearly uniform everywhere, and its composition (mostly hydrogen) well mixed. The entropy per unit mass of such a tenuous cloud is adequately given by the ideal gas formula,

$$S = C \ln T + R_g \ln V ,$$

where the previously undefined terms are C, the heat capacity mole$^{-1}$ ($\frac{3}{2}R_g$ for a monatomic gas), $R_g$, the ideal gas constant ($8.3 \times 10^7$ erg K$^{-1}$ mole$^{-1}$), and V, the volume of the cloud.

Now suppose a gravitational instability occurs. This change might be triggered by the passage of a parent galaxy's spiral-density wave, by the concussion induced by a supernova explosion of a nearby star, or by any number of other events (even solely chance perturbations in density could do it) that cause the interstellar cloud to begin infalling. The cloud's gravitational potential energy is thereby converted to kinetic energy, half of which is radiated away and the other half retained as heat (as stipulated by the so-called virial theorem of gravitational physics). That such a condensation results in a decrease in system entropy can be proved by calculating the entropy change per unit mass between the initial ("sub i") interstellar cloud and the finally ("sub f") formed star. In the previous notation, this is $\delta S_{sys}$, the entropy change produced by irreversible processes within the stellar open system:

$$\delta S_{star} = \delta S_{sys} = \frac{3R}{2m} \ln(T_f/T_i) + \frac{R}{m} \ln(\rho_i/\rho_f).$$

Here, m is the molar mass of atomic hydrogen and we have used the fact that $V_f/V_i = \rho_i/\rho_f$, where the densities $\rho_i = 10^{-22}$ g cm$^{-3}$ and $\rho_f = 1.5$ g cm$^{-3}$. Substituting also for $T_i = 15$ K and $T_f = 10^7$ K, the minimum temperature needed for hydrogen fusion, we find a net reduction in entropy per unit mass for the newly formed star:

$$\delta S_{star} = \delta S_{sys} \simeq -2.5 \times 10^9 \text{ erg g}^{-1} \text{ K}^{-1}.$$

According to the second law of thermodynamics, such an entropic decrease can occur only if there were an equivalent or larger increase in entropy elsewhere in the Universe. Computing the amount of entropy radiated away electromagnetically by the forming star, we find the increase in entropy of the surrounding interstellar environment per unit mass of material condensed is

$$\delta S_{env} = \delta Q/T = \delta U/M <T>,$$

where $\delta U$ is the change in gravitational potential energy during the star-forming process, M is the total mass of the star, and $<T>$ is the mean temperature at which radiation is released during the formative stage (namely, at which the original gravitational potential energy is dissipated). With spherical symmetry assumed for the original galactic cloud and a radius, r, for the final star, the gravitational potential energy, $U = -3GM^2/5r$, half of which is lost (i.e., $-\frac{1}{2}$) via radiative dissipation during the star-forming process. Finally, using typical solar values of $M = 2 \times 10^{33}$ g, $r = 7 \times 10^{10}$ cm, and $<T> = 1000$ K as the mean effective radiating temperature during formation, we have

$$\delta S_{env} \simeq +4 \times 10^{11} \text{ erg g}^{-1} \text{ K}^{-1}.$$

Thus, the increase in entropy of the surrounding interstellar environment is more than a hundred times greater than the computed decrease in entropy of the newly structured star (and plenty to overcome any unreasonable estimates in the above calculations). Clearly, the star has become an ordered clump of matter at the considerable expense of the rest of the Universe. The order is generated by the long-range, attractive gravitational force; the disorder is generated in the conversion of gravitational to heat energy, much of which the protostar ejects into its surroundings. The net effect of the *entire transaction*—the star's open system plus the surrounding environment—is a demonstrable increase in entropy, in complete accord with the second law of thermodynamics. We shall return to the issue of increased ordering within evolved stars in the Discussion.

## The Case of Cell Metabolism

Inanimate, physical systems are not the only ordered structures that experience a decrease in entropy, yet at a cost to their surrounding environment. All animate, biological systems do so as well. To be sure, living systems also enjoy dynamic steady-states that are neither stable nor static. For instance, our human bodies today are not the same ones we had seven years ago; hardly a cell is still alive that was part of that older body. More than just blood, flesh, and guts, even our skeletal parts are always changing as bone cells process new materials (air, water, and nutrients) to stay alive. The Russian biochemist Alexander Oparin had a nice analogy for this kind of ongoing flux in seemingly fixed entities. We are like buckets of water, he said, with water pouring in through a tap at the top and leaking out at the same rate through another tap in the bottom. The water in the bucket stays level and its outward appearance is constant, but its contents are regularly changing, the whole apparatus being quite unlike an ordinary bucket standing full of water. This is not to say that we resemble buckets of water, even sophisticated ones with lots of plumbing, for any form of life is demonstrably more complex than any abiotic system.

As a small example of cellular dynamics and its adherence to broad thermodynamic laws, consider some of the basic metabolic processes in living systems, such as food digestion, biochemical synthesis, and muscular contraction, each of these actions resembling the activity of a chemical engine more than a heat engine of conventional thermodynamics. Here, chemical energy need not be first converted to thermal energy to power the engine, as in a battery; with living organisms, chemical energy can be converted *directly* into mechanical energy, an essential feature of life itself. Perhaps we should more properly speak of "chemodynamics" in describing life, given that cells are essentially isothermal systems lacking temperature gradients and therefore unable to be powered by heat alone. Alas, rather than inventing yet more specialized lingo we shall instead embrace thermodynamics "with an asterisk," knowing that biochemists often use inclusive, non-equilibrium thermodynamic principles when modeling life, and especially its energy flow.

Even so, an intermediate step is required to power all life—bacteria, plants, and animals—suggesting (owing to its commonality) that this advance must have evolved early on in the history of life. As living systems metabolize their food, such as sugars and other carbohydrates, they produce a sort of fuel, called ATP, the molecule adenosine triphosphate. This chemical acts as a carrier of energy from the site in a body where food is digested to the site where it is used. This is the molecule that permits an organism to do work, mostly because its chemical bonds are rich in energy, a little like pent-up springs ready to unwind on behalf of the organism (Lehninger 1975).

Depicted in Figure 15(a), the adenosine molecule comprises adenine (a basic compound of carbon, oxygen, hydrogen, and nitrogen) and ribose (a sugar, or simple carbohydrate). Phosphate groups (a phosphorous atom with associated oxygen atoms) can attach to the ribose part of adenosine: three such groups for ATP, two and one for ADP and AMP, respectively, its diphosphate and monophosphate derivatives. It is ATP that acts as the primary agent to power work among all forms of life on planet Earth. In brief, here's how it does it.

When chemicals are taken into a living system—for humans, for example, in the form of meat, cheese, beans, and so on—the animal or vegetable proteins are broken down by digestive enzymes (catalysts) into some of the basic amino acids. This is an entropy-increasing process because somewhat ordered, large molecules are converted into many smaller ones having more randomized spatial arrangements and increased freedom of motion. We humans, among many other life forms, then synthesize new protein, combining the amino acids in the correct order and type to be useful to us. This is an entropy-decreasing process, which would not occur without the compensating, entropy-increasing combustion of fuel noted above. Let us tally the balance sheet for a part of the biological metabolism running the chemical engine of a living organism with a simplified flow diagram, shown in Figure 15(b).

Examine briefly a single reaction system that is representative of ATP participation in cellular metabolism, namely the assembly of the chemical glycogen, a long chain of hundreds and sometimes thousands of glucose molecules formed by photosynthesis and linked together as a polymer macromolecule; this is where animals store their carbohydrate (hence energy) supplies, most of it concentrated in the liver and muscles. Glucose is a ring-shaped sugar molecule having two parts hydrogen (H) for each carbon (C) and oxygen (O)—specifically, $C_6H_{12}O_6$. Chemical analysis (Stryer 1988) shows that the simplest possible case of linking only two glucose molecules yields a net decrease in entropy, for this is a process that builds up order:

$$C_6H_{12}O_6 + C_6H_{12}O_6 + \text{(free) energy} \rightarrow \text{glycogen,}$$
$$\delta S_{sys} = -5.2 \times 10^8 \text{ ergs K}^{-1} \text{mole}^{-1}.$$

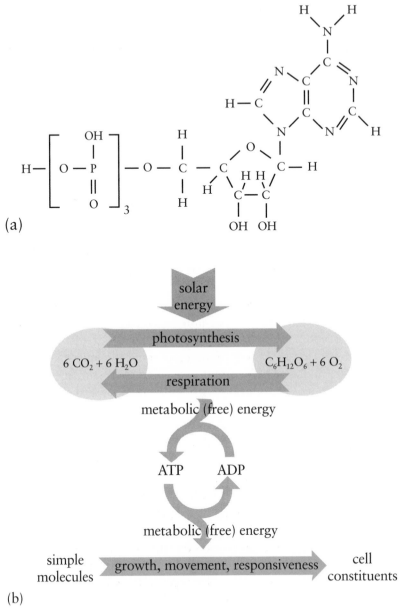

(a)

(b)

The free energy ($+3.8 \times 10^{-13}$ erg, or $+0.24$ eV) driving this endergonic synthesis comes from the exergonic conversion of an ATP molecule to ADP for each glucose added to glycogen; reacting spontaneously with a single water molecule ("hydrolysis"), ATP's energy-rich bond is broken with its terminal phosphate group ($PO_3H$), thereby forming a more stable system that releases the energy ($-4.8 \times 10^{-13}$ erg, or $-0.30$ eV) needed to power the above vital biochemical reaction within the body. And here, laboratory studies show that entropy is increased, for this is a process that breaks down order:

$$ATP + H_2O \rightarrow ADP + H_3PO_4 + \text{(free) energy},$$
$$\delta S_{env} = +9.1 \times 10^8 \, \text{ergs K}^{-1} \text{mole}^{-1}.$$

In this way, ATP acts as an intermediate fuel, helping to make large carbohydrate molecules from smaller ones, indeed to convert simple molecules into more complex cell constituents. The two-step process outlined here is a simplified version of the actual reaction-catalyzed pathways involving ATP, yet the initial reactants and final products as well as the numerical entropy gain for the real transaction are exactly the same. Biosynthesis of this sort resembles the formation and folding of a protein by the clustering of amino acids or the spontaneous assembly of a ribosome, both of whose increased organization is more than offset by the decrease in the order of the surrounding water molecules. Not surprisingly, the net effect of all these coupled biochemical reactions is an entropy increase, as once again required by the second law.

Much as for the increased ordering of physical systems (such as evolved stars), we shall return in the Discussion to consider once more

---

Figure 15. (a) Schematic diagram of the ATP molecule (adenosine triphosphate, $C_{10}H_{12}N_5O_4[PO_2OH]_3H$), showing its three phosphate groups ($HPO_3$, a phosphorous atom surrounded by oxygen atoms) attached to the sugar ribose (a carbohydrate molecule at center) and the organic-rich adenine compound (at upper right). The energy-rich chemical bonds of ATP comprise a kind of intermediate fuel, working for the body to synthesize order within a living organism—a carrier of energy and a producer of entropy, yet one that accords well with the second law of thermodynamics. The discussion in this section is only a small part of a larger process wherein many coupled reactions occur in a typical animal cell's metabolism. The schematic diagram in (b), still much simplified, shows the larger metabolic cycle of free-energy changes (from top to bottom) that ultimately, via respiration, converts simple molecules into more complex cell constituents.

the evolution of biological and cultural systems, all of whose complexifying inevitably and irreversibly causes the cosmic net entropy to rise. Such is the nature of change: The emergence of order and the growth of complexity, everywhere and on all scales, do exact a toll—and that toll means a Universe sinking further into an ever-disordered realm of true chaos.

## Summary of the Nature of Change

Nature's many ordered systems can now be regarded as intricately complex structures evolving through a series of instabilities. In the neighborhood of a stable (equilibrium) regime, evolution is sluggish or nonexistent because small fluctuations are continually damped; *destruction* of structure is the typical behavior wherein disorder rules. By contrast, near a transition (energy) threshold, evolution accelerates and the final state depends on the probability of creating a fluctuation of a given type. Once this probability becomes appreciable, the system eventually reaches a unique though dynamic steady state, in which *construction* of structure wherein order rules is distinctly possible. Such states are thereafter starting points for further evolution to other states sometimes characterized by even greater order and complexity.

Observations of evident and rising order at localized sites in the cosmos need not be seen as conflicting with the central tenet of modern thermodynamics, which stipulates disorder to be universally increasing. In particular, two different sets of physical laws are unnecessary to account for such diametrically opposite behavior in Nature. Researchers have now reached a general scientific consensus that the evolution of order is governed largely by a single physical principle, though one operating in demonstrably different physical realms: near and far from equilibrium. That single, unifying principle encompassing all aspects of natural change is the concept of energy flow as guided by the second law of thermodynamics.

# CHAPTER ONE

# MATTER

A lthough modern cosmology—Nature on the grandest scale—implies that matter only later emerged from the radiation of the early
Universe, it is pedagogically useful to quantify first the role of matter
and thereafter the primacy of radiation. In this way, the greatest change
in the history of the Universe—the transformation from radiation to
matter—can be clearly and mathematically justified. This preeminent
event is then contrasted with a second great change of cosmic evolution—the change from matter to life—in the third and last chapter of
this book.

At the outset of this chapter on "matter," let's define it. Matter is
anything that occupies space and has mass. All physicists the world
over agree on this simple definition. Alternatively, we could say that
matter is the stuff that is common to all material things. Thus, solids,
liquids, and gases are matter, but ideas, space, and photons are not.

### Galaxy Recession

Galaxies provide essential clues regarding the bulk nature of matter in
the Universe; they are the "test particles" that assay the texture of space
and time. As illustrated in Figure 16, images of distant galaxies also reveal a rich display of spectroscopic features from which both chemical
information and physical dynamics can be derived. The cosmologically
significant observations are embodied within Edwin Hubble's law, an
empirical finding named after the American astronomer often credited

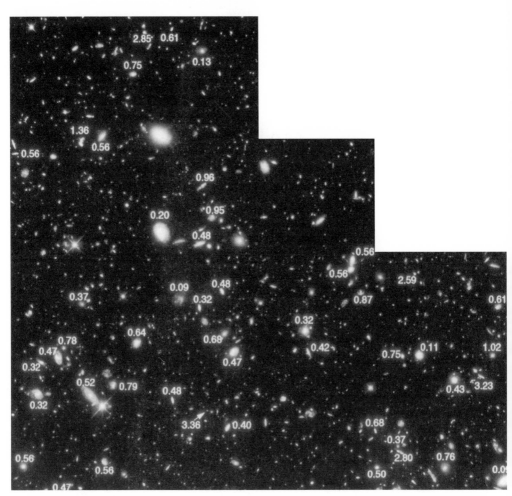

Figure 16. This is the most detailed view of faint objects in extragalactic space, called the Hubble Deep Field. An approximately 100-hour time exposure taken with the Earth-orbiting *Hubble Space Telescope,* it shows a fabulously rich array of thousands of galaxies in a minute area of the sky near the Big Dipper constellation. The numbers marked on the image are red shifts, z, measured spectroscopically by the Keck Telescope in Hawaii and interpreted as a Doppler effect. However complex cosmic structure may seem, including odd lace and peculiar filaments on even larger scales than shown here, that structure is simpler than that of the most primitive microbe (courtesy Space Telescope Science Institute).

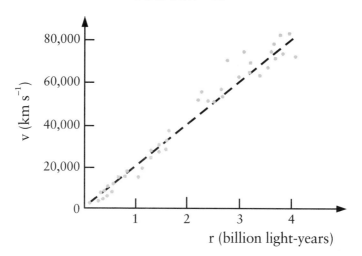

Figure 17. Data for a few dozen relatively nearby galaxies, whose recessional velocities are highly correlated with their distances, imply that galaxies and their related galaxy clusters have some definite organized motion in space. This is Hubble's law, and the dashed, least-squares fit to those data (the slope) is Hubble's constant, $H_0$ (adapted from Chaisson and McMillan 2000).

with the 1929 discovery of a correlation between two of the most basic properties among galaxies. (Here the word "law" is used advisedly, for this correlation is strictly an empirical result not based on any firm physical principle.) Figure 17 illustrates the essence of this relationship, where the data represent velocities and distances observed for numerous galaxies and the dashed line comprises the best linear fit to those data. Inherent within this law is the widely accepted interpretation that the observed red shifts of the galaxies' spectral lines are due to the Doppler effect—namely, that red shift is a velocity indicator.

Hubble's law expresses the observational finding that the radial velocity, v, with which a galaxy recedes is linearly proportional to its distance, r, from us. This law compactly describes the observed fact that the distances separating the galaxies, or more precisely the clusters of galaxies, are increasing with time; the finite velocity of light guarantees that looking out into space is equivalent to probing back into time. Virtually all galaxies obey this law—"virtually" because a few *nearby* galaxies, including the neighboring Andromeda Galaxy, are known to have a component of their velocity directed toward us. This is attributed to the random, small-scale motions that all galaxies display (generally within their parent galaxy clusters) in addition to their large-scale recessional motions as part of the "Hubble flow."

Symbolically, we write Hubble's law as

$$v \propto r ,$$

which becomes, after the insertion of a proportionality constant equal to the slope of the dashed line in Figure 17, a number generally known as Hubble's constant, H, a measure of the rate of galaxy recession,

$$v = H\, r .$$

Based on the best observations to date, $H_0$, the value of Hubble's constant today (hence the subscript zero and pronounced "H-naught") equals $65 \pm 20$ km s$^{-1}$ Mpc$^{-1}$; thus the galaxies recede with an additional velocity of 65 km s$^{-1}$ for every additional million parsecs of distance from us. (The "parsec" is a unit of distance at which a point source subtends an Earth-orbit parallax of 1 arc second; it equals approximately $3.1 \times 10^{18}$ cm, or 3.26 light-years.) The error, $\pm 20$ km s$^{-1}$ Mpc$^{-1}$, arises mostly from inaccuracies in the measurements of intrinsic brightness of some of the more distant and hence faint galaxies, making estimates of their distances a difficult task; by contrast, the velocities of remote galaxies are well measured from the red shifts of their various spectral features.

So, the greater the distance to a galaxy, the faster that galaxy recedes, much like expelled fragments of a huge detonation; stated more correctly, the galaxies and space itself are together receding. Presumably, the fastest-moving galaxies are by now farther away *because* of their high velocities. Visualizing the past by mentally reversing the outward flow of galaxies, we reason that all such galaxies were once members of a smaller, denser, and hotter Universe. Accordingly, we surmise that an explosion of cosmic proportions—popularly termed by some the "big bang" and by others "creation"[1]—probably occurred at some time in the remote past. The galaxies, including our own Milky Way, share in the expansive aftermath of this cosmic bomb, for they delineate, at one and the same time, both the underlying fabric of the Universe and the scattered debris of that primeval explosion.

Assuming (as do virtually all contemporary astronomers) that the recessional motions of the galaxies delineate the expansion of the Universe, we can use Hubble's law to derive the time elapsed since this grandest of all explosions began—namely, the age of the Universe. To do this, recall from elementary physics that r = vt, which when substituted into Hubble's law, reduces to "the Hubble time,"

$$t = H^{-1} .$$

Using the observationally estimated value of $H_0$, we find t = (15 ± 4) × $10^9$ years. This is the simplest possible estimate of the current age of the Universe and it would be the correct age if cosmic expansion had been uniform since the big bang. Actually, it is probably an upper limit because we have neglected cosmic deceleration (or acceleration) owing to the accumulated mass of the Universe, a correction that will be approximated later in this chapter. For now, we take the Universe to be 10–20 billion years old.[2]

Recessional motions of the galaxies comprise our best evidence that the entire Universe is active. Like everything within it, the Universe itself changes with time. To be sure, the cosmos is expanding in a directed fashion; in short, it is evolving.

## Cosmological Models

Now, leave these observational results for a moment and consider a theoretical argument. Imagine an arbitrary sphere of mass, M, and radius, r, expanding isotropically with the Universe at a velocity, v, from some central point, as shown in Figure 18. This sphere is not necessarily meant to represent the entire Universe as much as an exceedingly large, isotropic gas cloud, in fact larger than the extent of a typical galaxy supercluster (~100 Mpc across), which comprises the topmost rung in the known hierarchy of matter assemblages in the Universe. Nonetheless, this spherical system can be considered to mimic (and thus to model) the bulk dynamical behavior of the Universe as a whole.

Invoking the central dogma of modern physics—the conservation of energy—we set equal to a constant the sum (per unit mass) of the kinetic and potential energies of a typical galaxy on the sphere's surface—our cosmological "test particle"—namely,

$$v^2/2 - G\,M/r = constant .$$

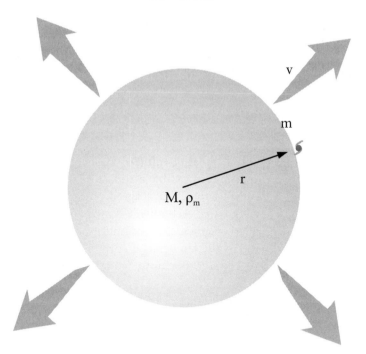

Figure 18. A typical galaxy of mass, m, is imagined here on the surface of an isotropically expanding sphere of radius, r. The sphere's contained mass, M, is distributed uniformly with density, $\rho_m$, yet gravitationally acts as though it resides in a single point at the sphere's center.

Here, G is the universal gravitational constant equal to $6.67 \times 10^{-8}$ dyne-cm$^2$ g$^{-2}$, and the total mass, M, of the matter within the sphere of average density, $\rho_m$, is given by

$$M = \tfrac{4}{3} \pi r^3 \rho_m .$$

The minus sign in the middle term of the energy-conservation equation conforms to the common convention in physics that gravitational potential energies are inherently negative; kinetic energy can increase only at the expense of potential energy. Mass beyond the sphere need not be considered, as the so-called Birkhoff theorem states that matter outside the sphere has no gravitational effect on the behavior of what is inside.

For this Newtonian argument to be consistent with relativistic cosmology, the constant term on the right-hand side is often taken to equal $-kr_0^2/2$, where k is a time-independent curvature constant related to the total energy of the system sketched in Figure 18 and $r_0$ is the current radius of that system (i.e., at the present epoch). Newton's classical theory of gravitation not only simplifies our arguments, but also, for spherically symmetric mass distributions (like the one considered here), yields many important results that are essentially the same as those derived from Einstein's theory of relativity. With the exception of the propagation of radiation, general relativity plays a surprisingly small role in many of the most interesting cosmological applications. Although relativity is needed to explore the often peculiar *geometry* of space on large scales, Newtonian physics is a perfectly adequate description of the global *dynamics* of a changing Universe. The curvature constant, k, is conventionally scaled so as to have the values 0 (flat Euclidean geometry), or ±1 (positively curved Riemannian or negatively curved Lobachevsky geometries). Substitution of Hubble's law and of the above expression for the total mass rearranges our energy-conservation equation, to wit,

$$H^2 r^2 - \tfrac{8}{3} \pi \, G \, r^2 \, \rho_m = -k \, r_0^{\,2}.$$

At this point in cosmological models, astrophysicists often introduce a dimensionless, time-dependent term called the "universal scale factor," R, which relates the radius, r, at any time, t, in cosmic history to the current radius, $r_0$, at the present epoch, namely,

$$r = R \, r_0.$$

The scale factor, R, is a measure of the overall expansion of the Universe as a function of time, and, even in the case of an infinite Universe where "radius" means little, can be thought of as typifying the average distance separating clusters of galaxies. This equation is true only because the system of Figure 18 was hypothesized to be an arbitrary sphere, and so its equations must apply to all possible spheres containing fixed amounts of matter and therefore to all particles in the Universe. The assumption that the Universe is both homogeneous everywhere and isotropic about every point within it, formally called the cosmological principle, is backed by observations on scales larger than that of galaxy superclusters, especially that of the cosmic background radiation discussed in Chapter Two. Though its strict validity is unknown, this principle surely does simplify the study of cosmology, in addition to providing a universal reference frame against which to measure common, cosmic time, as noted earlier. Thus, R = 1 and r = r$_0$ at the current epoch, whereas, for example, in the early Universe, R << 1.

By inserting the scale factor into the above equation for energy conservation, we arrive at one of the most basic relations of modern cosmology—the Friedmann-Lemaitre equation, named after the Russian meteorologist and the Belgian cleric who in the 1920s independently formulated the problem as we now know it:

$$H^2 - \tfrac{8}{3} \pi \, G \, \rho_m = -k \, R^{-2} \, (+ \Lambda).$$

The so-called cosmological constant, $\Lambda$, is added here to denote an hypothesized repulsive term (a kind of anti-gravity) that would act on matter only on the largest scales, a peculiar quantity whose physical significance is mostly a mystery and numerical value largely unknown. $\Lambda$ has units of inverse time squared, and its associated force of repulsion at all points in space—a wholly new fifth force in the Universe and of magnitude $\Lambda r$—would grow larger with time, thereby escalating to runaway expansion on large scales, all the while having negligible effect on small scales so as to avoid interfering seriously with the law of gravity so well tested locally in the Solar System. The $\Lambda$-term is noted here partly as a hedge and partly as a courtesy to colleagues who feel it could play a role, but I put it in parentheses to denote my stance that it is not likely a factor in cosmic evolution given that the Universe is yet too young. Chalk it up, either way, to metaphysical posturing. This peculiar term was conceived by Einstein in 1917 in order to keep his universe model from expanding (or contracting); like most other leading scientists of the day, Einstein was an Aristotelian at heart, preferring to believe that the Universe ought to be static and unchanging. When Hubble and colleagues a decade later made clear the recession of the galaxies and its implication for an expanding Universe, Einstein removed the term, admitting that it was a big mistake. It resurfaced again in the 1950s when proponents of the now-discredited steady-state model of an eternally static Universe needed a "fudge factor" to, again, keep things steady on the largest scales. Periodically, $\Lambda$ comes and goes in cosmological circles, usually gaining temporary favor whenever some of the key numbers sought by cosmologists, such as the Hubble constant, the age of the Universe, or the average density of matter, seem not to be mutually consistent. And then it quickly disappears for decades. Observers have not been able to find compelling evidence for it, and theorists have not been able to absolutely require it or even

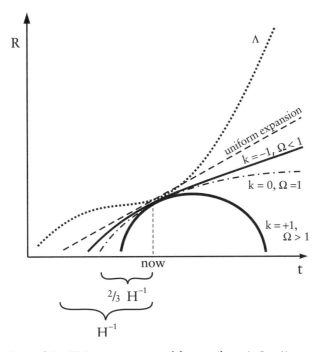

Figure 19. An evolving Universe can expand forever (k = −1, Ω < 1) or contract back (k = +1, Ω > 1) to a point much like that from which it began. The parabolic solution to the Friedmann-Lemaitre equation (k = 0, Ω = 1) mathematically stipulates that the Universe is barely open, reaching infinity with a zero velocity (meaning that it expands forever), whereas the hyperbolic solution (k = −1) implies that the Universe is demonstrably open, arriving at infinity with a finite velocity and "moves beyond" (meaning that it also expands forever). For the closed model the age of the Universe is less than $\frac{2}{3}H^{-1}$, whereas for an open model its age is between $\frac{2}{3}H^{-1}$ and $H^{-1}$. The dotted curve, labeled Λ, shows the effect of adding a positive cosmological constant, which makes the age of the Universe indeterminate; it produces a steady "coasting phase" (at left) wherein the Universe ages without much expanding, followed by an exponentially accelerating expansion (at right) to infinity, regardless of the cosmic mass density.

explain it. Early in the twenty-first century, cosmologists' interest once again has been piqued regarding the cosmological constant based on observations of distant supernovae which imply, surprisingly, that the Universe is accelerating. In this book, however, we shall take Λ = 0, although occasionally note the consequences should Λ be finite.[3]

Figure 19 sketches various solutions to the Friedmann-Lemaitre equation. Most such evolutionary models decelerate, and thus their solutions curve concave toward the time axis, in contrast to the idealized (matterless) model of uniform expansion, also graphed as a dashed line in Figure 19. The Friedmann-Lemaitre solutions illustrate that the Universe can be "open" (i.e., $k = -1$) and therefore recede forevermore to infinity, or "closed" (i.e., $k = +1$), in which case its contents eventually stop receding and thereafter contract to a point presumably much like that from which the Universe began. The intermediate case (i.e., $k = 0$) corresponds to a Universe precisely balanced between the open and closed models; in fact such a model Universe would eternally expand toward infinity and never contract, and is thus also open. (Figure 19 also includes a typical solution to the above equation with a finite cosmological constant, $\Lambda$. Note how the future evolution of R is quite different, even allowing for universal acceleration *ad infinitum,* owing to $\Lambda$'s repulsive nature on very large scales. By contrast, the past evolution of R is not so different from any of the other models, implying that $\Lambda$, even if real, has had little influence on cosmic evolution to date.)

### Two Useful Notes

Before proceeding with our quantification of matter's evolution, note two consequences of the scaling factor that will be useful later in this chapter. First, the equation for the scaling factor proves that Hubble's constant is not really a constant; like virtually everything else in the Universe, it changes with time. To see this, take the time derivative of the above equation that introduced the scaling factor, namely,

$$dr/dt = v = r\,R^{-1}\,dR/dt \,,$$

or

$$H = R^{-1}\,dR/dt \,,$$

which means that

$$H \simeq t^{-1} \,.$$

Accordingly, Hubble's constant was larger long ago and will be smaller in the future, as expected for the expansion parameter of any explosion, big-bang or otherwise; that is also why we use $H_0$ to denote the current value of H for the present, or "zeroth," epoch.

Second, note that the Doppler shift, z (which is used to derive galaxy velocities, namely, $z = v/c$ for $v \ll c$, where c symbolizes the velocity of light and equals $3 \times 10^{10}$ cm s$^{-1}$), is defined as the difference between the actual observed wavelength, $\lambda_{obs}$, and a standard laboratory wavelength, $\lambda_{lab}$ (in a rest frame, in this case collocated with the cosmic object), all normalized by $\lambda_{lab}$. Recessional velocities and hence red shifts are conventionally taken to be positive and conversely, thus,[4]

$$z = ( \lambda_{obs} - \lambda_{lab} )/\lambda_{lab} .$$

Local galaxies are receding slowly and have small values of z, some quite close to zero; the most distant galaxies—many of them so-called quasars—are rapidly receding and thus have higher values of z, although none currently observed (and confirmed) to exceed $z = 6$. Taking both waves (photons of radiation) and particles (of matter) to be conserved in our modeled sphere (thus neglecting absorption and emission events), and noting that the wavelength of radiation changes in precisely the same way as the radius of the sphere, we can use the scaling factor to recast the above equation,

$$z = R^{-1} - 1 ,$$

or as is often used in the research literature,

$$(1 + z) = R^{-1} .$$

Thus, R varies inversely as z, implying, for example at $z = 1$ and $R = \frac{1}{2}$, that the size of the Universe was half that at present and that the "look-back time" was approximately 4 billion years after the alpha point at the start of all things (which occurred at $z = \infty$ and $R = 0$).

### Evolution of Cosmic Density

To follow the evolution of matter throughout cosmic history, return to the Friedmann-Lemaitre equation. Note that this equation is clearly simplest when $k = 0$—the so-called Einstein-deSitter case, named after the two scientists who derived its mathematical solution in 1932—for then it reduces to exactly

$$\rho_{m,c} = 3 H^2/8 \pi G .$$

We work with the $k = 0$ model henceforth partly out of simplicity and partly to concede the possibility of an important yet unconfirmed event—"inflation" in the very early Universe—that demands k = 0 (Guth 1997; see also Chapter Two, note 3). Models with k ≠ 0 are much more difficult to solve yet would not change appreciably the main lines of argument in this book. Evaluating for G and for a value of $H_0 = 65$ km s$^{-1}$ Mpc$^{-1}$, we find that $\rho_{m,c} = 10^{-29}$ g cm$^{-3} \simeq 6 \times 10^{-6}$ atom cm$^{-3}$. This extremely thin spread of matter, roughly 6 atoms per cubic meter or about a million times more rarefied than the matter in the "empty space" between Earth and the Moon, equals the "critical density" (hence the extra subscript "c") above which the Universe is closed and below which it is open.

To further stress the relationships among the matter density, the spatial curvature, and the long-term evolution of the Universe, and to connect with standard cosmology literature addressing ultimate ends, we introduce here one more symbol (appropriate for the subject), $\Omega$, the "universal density parameter." This is the ratio of the actual density of the Universe to its critical value for closure, namely,

$$\Omega = \rho_m/\rho_{m,c} \, .$$

Therefore, for the special, zero-curvature case, k = 0, the two densities are precisely equal and $\Omega = 1$; for the elliptical curvature case, k = +1, so $\Omega > 1$, meaning that the Universe collapses back to a point, whereas for the hyperbolic curvature case, k = −1, $\Omega < 1$, and the Universe expands forevermore. (The relationship between k and $\Omega$ is more complicated, as for much of the rest of cosmology, if the cosmological constant, $\Lambda$, is non-zero.)

Whether the actual, current density is smaller or larger than the critical value, making the Universe open or closed, respectively, is currently unknown. At face value, by counting merely the number of galaxies whose light is observed within a typical galaxy cluster, attributing to each of them an average galaxy mass, and dividing by the volume of the cluster (whose distance needs to be known), the current density, $\rho_{m,0} \simeq 10^{-31}$–$10^{-30}$ g cm$^{-3}$, implying $\Omega \simeq 0.01$–$0.1$ and an open Universe. However, the fate of the Universe remains unresolved, largely because of the uncertainty concerning the extent of "dark matter" within and around galaxies. Such dark matter is inferred to lurk in the halos of most galaxies, based on observations of the dynamics of outlying stars, and also to be present as "intracluster gas" within galaxy clusters, based on the amount of mass needed, but not found, to bind the clusters into their observed groupings. In any case, as regards cosmic evolution's broad sweep, and aside from issues of ultimate fate, the amount of dark matter is hardly crucial; where dark matter would make a difference, provision is made for it in the ensuing calculations without much further comment on its elusive nature. Flatfootedly stated though no less true, the origin and destiny of the Universe—subjects for which there are little data—are hardly relevant to the scenario of cosmic evolution; we are concerned in this book mainly with the many varied events that occurred once the Universe began and thereafter to the present. To be sure, cosmic evolution would still approximate reality as an historical narrative or evolutionary epic even if a deistic God put it all in motion at $t = 0$. See Coles and Ellis (1997) for a critical review of the issue of dark matter and how it might affect the fate of the Universe.

An interesting point here is that the $k = 0$, $\Omega = 1$ case is a good representation of events in the early Universe, regardless of whether the Universe is open or closed. To see this, examine the complete Friedmann-Lemaitre equation and note that, for small R (that is, for early times), the right-hand term, which scales as $R^{-2}$, will always be negligible compared to the left-hand term, which scales as $R^{-3}$, thus guaranteeing the $k = 0$ solution as a reasonable approximation whenever $r \ll r_0$. This point is also made clear by studying the curves of Figure 19; for the early Universe, the family of curves for all k nearly overlap, meaning that virtually any value of k suffices when $R \ll 1$.

Let us formalize one more conservation principle, namely, that the number of particles of matter remain constant within the sphere of Figure 18, so that as the Universe expands, its density naturally grows smaller. Taking $\rho_{m,0}$ as the current matter density of the Universe, we can express this conservation principle simply:

$$\rho_m = \rho_{m,0}\, R^{-3} .$$

This is just as expected intuitively since the volume of a sphere scales as $R^3$. This particle-conservation equation is valid except in the very earliest moments of the Universe, when many particles were themselves only materializing from primordial energy at $t < 1$ second, a time frame of minor concern here given the "big-picture" posture of this book.

By substituting the above particle conservation equation into the simplest $k = 0$ case of the Friedmann-Lemaitre equation, we then find

$$H^2 = 8\,\pi\,G\,\rho_{m,c}\, /\, 3\,R^3 .$$

And recalling our note regarding the variability of Hubble's "constant," we can manipulate this equation to show explicitly how the matter density changes as a function of time throughout the history of the Universe,

$$\int dt = (\tfrac{8}{3}\,\pi\,G\,\rho_{m,c})^{-0.5} \int R^{0.5}\,dR\,.$$

Integrating, we find

$$t = \tfrac{2}{3}\,H^{-1}\,,$$

where the $\tfrac{2}{3}$ factor accounts for the cosmic deceleration neglected at the beginning of this chapter, but noted in the Figure 19 caption. Thus, for the special ($k = 0$, $\Omega = 1$) Einstein-deSitter model, the present age of the Universe, $t_0$, could be as young as $10 \pm 2$ billion years. Using the earlier derived relation, $(1 + z) = R^{-1}$, we find two more useful expressions,

$$t = 3 \times 10^{17}\,(1 + z)^{-1.5}\ \text{[seconds]}$$

and

$$\rho_m = 10^{-29}\,(1 + z)^3\ [\text{g cm}^{-3}]\,,$$

both of which reduce to the aforementioned $10^{10}$ years and $10^{-29}$ g cm$^{-3}$ for the current epoch, $z = 0$. These last two equations can be further manipulated to show the temporal dependence of the matter density,

$$\rho_m \simeq 10^6\,t^{-2}\,,$$

where $\rho_m$ is expressed in g cm$^{-3}$ and $t$ in seconds.

Here, then, is a more technical interpretation of the principal curves in Figure 19: If $\Omega = 1$, then $R \sim t^{2/3}$, and the Universe expands gracefully at this same rate for all time. If $\Omega < 1$, $R \sim t^{2/3}$ also, but only until it reaches a fraction $\Omega$ of its present size, at which point it then enters free expansion with $R \sim t$. If $\Omega > 1$, the Universe expands more slowly than the special, intermediate case where $\Omega = 1$, and then collapses—a dire fate best not contemplated. Actually, as we shall show below, when radiation dominated matter in the early Universe, $t < 10^5$ years, $R \sim t^{1/2}$, so that all the Friedmann-Lemaitre curves would have behaved slightly differently than described until matter began dominating radiation (as it does now, and has since $t \sim 10^5$ years). However, given the scale of the abscissa, this slightly different temporal dependence of R would not

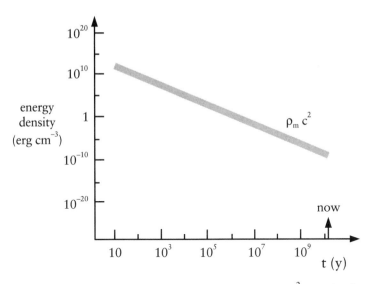

Figure 20. The temporal behavior of matter energy density, $\rho_m c^2$, on a log-log graph. The plotted line's width, or variance, represents the considerable range of uncertainty in the value of $\rho_{m,0}$ today; at issue is the amount of "dark matter" in the Universe, which is currently unknown.

even be noticeable so close to the ordinate. As for the curve labeled Λ in Figure 19, the effect of a cosmological constant is negligible at early (past) times, but dominates both radiation and matter at late (future) times.

We have therefore derived a way to quantify the evolution of the matter density, in bulk, throughout all of universal history. Hindsight suggests that it will be more useful to reexpress this quantity in terms of the equivalent *energy* density of that matter. We can do so by invoking the Einsteinian mass (m)-energy (E) relation, $E = mc^2$, that is, by multiplying the above equation for $\rho_m$ by $c^2$. Figure 20 illustrates the change of matter's equivalent energy density, $\rho_m c^2$, throughout the history of the Universe; we shall return to it in the next chapter in order to compare the evolution of matter's energy density with that of radiation's energy density.

## The Age Controversy

Debate continues to swirl around the vexing issue of the age of the Universe and of some of its contents. In brief, the problem is this: At face value, the Universe seems younger than some of the stars within it! Clearly, something is wrong, for no child can be older than its mother. At the most, our basic knowledge of physics or astronomy is incomplete; at the least, some slight adjustments need to be made in the analysis of the observational data. Chances are good that the latter pertains (Chaisson 1997).

As noted above, the age of a uniformly expanding Universe, $t = H_0^{-1}$, which for a popular, average value of $H_0$ (65 km s$^{-1}$ Mpc$^{-1}$) equals 15 billion years. This is a correct age only if the cosmic density is much lower than the critical value, $\rho_{m,c}$, in fact strictly correct only for an unreal Universe devoid of all matter. If, however, $\rho_m = \rho_{m,c}$, then the Universe decelerates with time and the true age must be less than 15 billion years; this is the special, $k = 0$, $\Omega = 1$ case of a perfectly balanced Universe, for which the solution was found in an earlier section of this chapter, namely, $t = \frac{2}{3} H_0^{-1} \simeq 10$ billion years. Since we also have some theoretical reasons (mostly having to do with the concept of inflation in the early Universe; see note 3 in Chapter Two) for supposing that the cosmic density is close to critical, astronomers suspect that the value of $H_0^{-1}$ might be an upper limit for the age of the Universe.

By contrast, a key component of the Universe seems older than 10 billion years. These are the ancient stars of the globular clusters, tight-knit groups of many thousands of stars strewn throughout the halos of galaxies that are probably as old as the galaxies themselves. Astronomers estimate such stellar ages based on the rates at which stars undergo nuclear fusion, a rather well-understood part of cosmic evolution. As depicted by the simplified diagram of Figure 21, the theory of stellar evolution specifies when bright blue, main-sequence stars should begin to change into smaller red giants; the fewer the number of bluish stars as compared with the number of reddish stars, the older the cluster (Kaler 1998). Many such globular clusters have been studied in this way and most of them show ages of $15 \pm 3$ billion years. Hence the paradox at hand: Some stars seem older than the Universe itself, a clear embarrassment to astronomy if not resolved.

Actually, this age problem is not new; it has existed in one form or another for well over a century, recurring periodically and then subsiding. For example, when, in the early nineteenth century, the pioneers of geochronology sought to assess the age of the Earth on grounds other than religious or philosophical, they essentially made two assumptions: that the Earth probably formed at the same time as the Sun, and that the Sun shone by the burning of some known chemical, like the wood

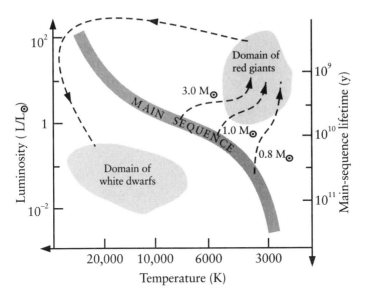

Figure 21. The subject of stellar evolution—one of the seven major phases of cosmic evolution—is fortunate to have a single graph that encapsulates many of the salient features in the "birth, life, and death" of a star. Known as the Hertzsprung-Russell diagram (after the Dutch and American astronomers who pioneered it), this graph organizes stars by luminosity (vertical axis) and surface temperature (horizontal axis). Several patterns and groups are evident: Most normal stars lie on the so-called main sequence, while red-giant stars are found at upper right and white-dwarf stars at lower left. Our Sun is currently near the middle of the main sequence, at 1 solar luminosity ($4 \times 10^{33}$ erg s$^{-1}$) and 6000 K surface temperature. The dashed lines depict the evolutionary tracks of 0.8, 1.0, and 3.0 $M_\odot$ stars, thereby dating the ages of star clusters depending upon their turnoff coordinates (marked in years) in the H-R diagram (Chaisson and McMillan 2000).

or coal commonly used during the Industrial Revolution. The answer they got for the age of the Sun, and hence the age of the Earth, was a few thousand years, a value less than that of recorded history. So an age controversy developed, not so much heated as merely amusing to most theologians of the time: How could Earth be younger than the duration of human existence?

The first assumption of early Victorian science was a good one (indeed we today consider the birth of Earth and Sun to be coterminous), but the second one was most definitely not; the Sun, assuredly, is not made of wood or coal. Lord Kelvin, von Helmholtz, and others later revised these calculations in the second half of the nineteenth century, taking the Sun to be made of an incandescent liquid mass (such as gasoline or kerosene) and allowing for some energy generation via gravitational infall (including meteors crashing into the Sun). They were still unable to increase the age estimate for the Sun to much more than 100 million years, a value surely older than recorded history, but much less than that indicated by the fossil record as interpreted by Darwin. At the time, long-dead life forms seemed at least several hundred million years old (and we now realize they are even older). Kelvin got similarly low values when trying to estimate the rate of cooling of Earth itself, partly because he overlooked thermal convection of the rocky interior, all of which put geological evolution into conflict with biological evolution. Thus, the age controversy continued, dominating scientific circles a century ago. Some of the debate (then as now) was quite vehement: How could life on Earth be older than the planet itself?

These early age discrepancies eventually went away. As radioactivity became better understood by Henri Becquerel, the Curies, and Lord Rutherford around the turn of the twentieth century, geologists could then measure the age of rocks directly. And what they found was a planet of a few billion years, which was then fully enough to provide the long time scales needed to explain Darwin's fossils. We now know that biological evolution has occurred over the course of some 3.5 billion years, and the Moon and meteorites, hence the Earth, are 4.6 billion years old.

Alas, in the 1930s, a version of this problem resurfaced. At issue then was Hubble's first measurement of $H_0$. Owing to observational uncertainties in the brightnesses of galaxies and especially to calibration errors in the analysis of the acquired data (from Cepheid variable stars), he and his colleagues found $H_0 \simeq 500$ km s$^{-1}$ Mpc$^{-1}$. This meant that $H_0^{-1} \simeq 1$–2 billion years for the age of the Universe and suddenly the general problem was back: How could Earth be older than the Universe?

In turn, this problem faded away over decades as many astronomers accumulated better observations and data analyses of the brightnesses and distances of the galaxies. By the 1950s, the value of $H_0$ had decreased fivefold and $H_0^{-1}$ had equally lengthened to about 10 billion years. Hence, the Universe was safely older than Earth, and the age problem disappeared . . . for a while. To be sure, it has resurfaced in more recent years, as noted above. Thus by the 1980s and into the 1990s, a modern version of the recurring age discrepancy has emerged:

How could the globular-cluster stars within the Milky Way be older than the Universe itself? Well, they can't be. It's as simple as that. Something is awry, again.

One possible solution to the current age dilemma invokes novel properties of the Universe, in effect "new physics." To give but one example, some researchers argue that the aforementioned "cosmological constant," or $\Lambda$-term in the Freidmann-Lemaitre equation would have the effect of putting the Universe into a "coasting phase," during which time it can grow older without expanding. Hence it is a useful mathematical term—some would say a weak crutch to force the Universe to be older than its constituents—and it is again gaining popularity in research circles. Figure 19 sketched how the Universe would evolve, in particular how R varies with t, with a typical $\Lambda$-term included. We do so for completeness, not because we think it is needed.

Another way out of the age discrepancy is to note that $H_0$, as currently measured, may not be an accurate indicator of the age of the Universe—whether it is $H_0^{-1}$, or $\frac{2}{3} H_0^{-1}$, or any other fraction of $H^{-1}$. Values of $H_0$, even those determined with today's best telescopes, might be "contaminated" by local velocity flows in the relatively nearby Universe; we might not be measuring the true expansion rate of the Universe. Galaxies beyond 200 Mpc are simply too dim to figure into the current search to specify $H_0$, and other large-scale motions inside this distant realm having little to do with universal expansion could be affecting even our best estimates of $H_0$. For example, we know that Earth orbits the Sun at a velocity of 30 km s$^{-1}$, the Sun rounds the Milky Way Galaxy at 220 km s$^{-1}$, the Galaxy pivots about the center of the Local Group of galaxies (toward the Andromeda Galaxy) at about 50 km s$^{-1}$, and the Local Group moves within the Virgo supercluster of galaxies at roughly 200 km s$^{-1}$. On larger scales, the Virgo supercluster is inferred to be moving at a velocity of another 400 km s$^{-1}$ toward an immense concentration of mass in the sky of the southern hemisphere, called by some the Great Attractor, but which is probably no more than an especially rich collection of several galaxy clusters some 150 Mpc away. Peculiar velocity flows on even larger scales are also possible, but not yet deciphered, in fact not even well probed at r > 200 Mpc. The point is that none of these "local" motions have much bearing on the largest movement of all—the Hubble flow, indicative of universal expansion. The above motions result from gravitational interactions on less-than-universal scales, and are independent of the grand expansion rate. Yet all these smaller motions are superposed on that larger flow, making tricky the derivation of $H_0$, based on observations of relatively nearby galaxies. This might be the origin of the apparent conflict between the age of the Universe and the age of the oldest stars within it. We simply might not have a good handle on $H_0$, as pertains to the overall Universe on a scale of 4000 Mpc, having measured it merely on a scale of 200 Mpc.[5]

The most likely outcome is that the current age controversy will simply fade away, just as better observations and improved data analyses caused similar glaring contradictions to evaporate during the past century. Indeed, several recent developments favor the dissolution of this problem altogether. For example, today's astronomers seem to be converging on smaller values of $H_0$, implying that the age of the Universe may have been recently underestimated; furthermore, unless more dark matter is actually found, then k could be distinctly less than unity and the Universe decidedly open, the further consequence being that the universal age is indeed closer to $H_0^{-1}$, the upper bound of which, say for $H_0 = 60$ km s$^{-1}$ Mpc$^{-1}$, would then accord with the lower bound of the most ancient stellar ages, namely about 12 billion years. What's more, recent reanalyses of some globular cluster data (especially enhanced helium abundances, which can raise the intrinsic luminosity of such stars, and results from the *Hipparcos* satellite, which revise upwards the brightnesses of several key variable stars) suggest that the globulars might have had their ages overestimated by nearly 20 percent. If true, then the above-noted age of 15 billion years for the oldest stars needs to be readjusted to about the 12-billion-year value toward which many current studies seem to be headed. That is the value used in this book, equivalent to some 12 million millennia.

For our purposes, we need not be overly concerned about the current age controversy, other than to note it as an active area of forefront research that seeks to specify a number (the value of $H_0$) to an accuracy of better than a factor of two, when many other cosmologically significant numbers (including those on which it depends, such as $\rho_{m,0}$) are known only to within a factor of about ten. Frankly, it is remarkable that the ages of the cosmos, stars, and life are so close and that together they seem to be stacking up so well along an ordered arrow of time. As for our cherished scenario of cosmic evolution, the subject is hardly affected by this lingering age controversy; the arrow of time itself can be contracted or expanded, a little like an accordion, to match the true age of the Universe, whatever it turns out to be. It is the sequence of events along the arrow that is more important than the magnitude of the arrow itself.

## Summary of Chapter One

Cosmology at the end of the twentieth century is both exhilarating and frustrating—exhilarating because much good data are now enriching a previously observationally starved subject poised at the threshold of answering some of the most fundamental questions, yet frustrating because some of the answers to those questions are eluding, even confusing, us. Some researchers find the age of the Universe problematic, others fret over the rate of cosmic expansion, while everyone seems bewildered regarding the nature and whereabouts of the dark matter. Fortunately, for purposes of articulating the broadest view of the biggest picture, none of these unsolved mysteries likely affect the main lines of argument in this book. The cosmic-evolutionary scenario cares little whether the true age of the Universe is 10 or 20 billion years, provided that the historical sequence of events is correct along time's arrow; likewise, whether the Universe decelerates, accelerates, or essentially expands at a constant rate is but a refinement for the cosmic evolutionist, provided that the Universe is indeed expanding at *some* rate sufficient to give rise to gradient-rich environments and thus to increasing amounts of complexity and intricacy during the course of universal history. Understanding the cause of the ongoing trend toward enhanced richness and diversity among Nature's organized systems is our task at hand, and, surprisingly, the answers seem largely independent of lingering cosmological puzzles.

As of 1999, the most distant known material object, a juvenile galaxy with the peculiar catalog name of 0140+326RD1, has a red shift of $z \simeq 5.4$, which according to the above equations, implies that $t \simeq 1$ billion years. This particular object will not last long as the record-holder, as photons now streaming toward our telescopes almost certainly include data on higher red-shifted objects. Because of the way that distance scales with z, however, distances to new objects will not be much greater than those already found, nor will any of the conclusions in this chapter likely change much at all. Thus, light from this remote object began its journey to Earth when the Universe was about 10 percent of its current age. This means that direct study of the farthest known matter can extend our knowledge back roughly 90 percent of the past, but that it cannot help us directly penetrate much closer to creation (at $t = 0$) than a billion years or so. All other known objects in the Universe are closer to us in space and therefore their signals were launched in more recent times. By contrast to material objects per se, only observation and analysis of pure cosmic radiation can unveil earlier epochs, thus getting us closer to the origin of everything.

# CHAPTER TWO

# RADIATION

The previous chapter noted how contemporary cosmology is guided largely by observation of material objects known as galaxies, which enable us to model the bulk change of matter now and at earlier epochs of the Universe. In this chapter, we consider another key observation, this one of a non-material invisible signal which, in turn, enables us to explore changes in radiation. Knowledge of the changes, or evolution, of these two quantities alone—matter and radiation—yields a veritable history, a rich natural epic of many majestic, temporal events and achievements through all time. Astronomy is as much of an historical subject as is the study of western civilization, of anthropology and archaeology, or even all of biology or geology. Astronomers are genuine historians; they go all the way back.

Radiation is primary in the study of the early Universe. It was only for the pedagogical reason that it is often more straightforward to analyze matter, and its evolution, that we initially undertook the study of matter in Chapter One. Radiation surely dominated the first great era in cosmic history, and it is only from changes in radiation that matter has emerged.

## Cosmic Background Radiation

Discovered serendipitously in 1965 by the American physicists Arno Penzias and Robert Wilson, weak, low-energy microwave radiation seems to flood the Universe, and sounds, to a radio astronomer, much like the hiss or static on an ordinary home (AM) radio receiver. When converted into a video signal capable of display on a computer terminal, this radiation resembles the "snow" seen on a television tuned to an inactive channel; a small part of that static is, in fact, due to the cosmic background radiation. Regardless of the direction observed or the time of day, night, or year, after all the familiar signals (such as well-known cosmic sources, atmospheric emissions, and man-made interference) are subtracted from the data, there remains a low-intensity, omnipresent microwave signal whose origin seems directly related to the aftermath of the birth of the Universe. This cosmic background radiation is presumed to be a remnant of the fiery origins of the Universe, a virtual relic or oldest "fossil" of the primeval explosion that commenced the cosmic expansion some 12 billion years ago.

Figure 22 summarizes the principal measurements of this radiation; the data points represent the observed radio and infrared intensities, $B_\lambda$ (in units of erg cm$^{-2}$ s$^{-1}$ steradian$^{-1}$ Hz$^{-1}$), measured at different wavelengths, $\lambda$, while the dashed curve comprises the best theoretical fit to those data. This curve, which peaks at $\lambda \simeq 1$ mm, is reproducible to very high accuracy in any direction of the cosmos; the observed radiation has been found to be remarkably isotropic to within one part in $10^5$, which further indicates that this ubiquitous radiation is of universal nature and not likely associated with a single material object or even a group of such objects. (Its intensity is at least ten times greater than all the stars, galaxies, and other radiating objects combined and averaged over the volume of the observable Universe.) This background radiation is also remarkably good confirmation of the cosmological principle, the central assumption of homogeneity and isotropy made when deriving the Universe models in Chapter One. Apparently all of space—even the otherwise totally empty intergalactic realms separating the galaxy clusters—is uniformly inundated with this "afterglow" perceived throughout the whole sky; the number density of photons, or the density of radiant energy, must be virtually the same everywhere.[1]

The dashed curve in Figure 22 does not have an arbitrary shape. In fact, it closely follows the blackbody spectrum of radiation emitted by any warm object whose elementary particles release energy while interacting with one another. Blackbody radiation is defined as radiation that has been thoroughly randomized through its interaction with matter, much as the radiation deep inside the cores of stars is completely randomized. The resulting thermally emitted radiation obeys the fol-

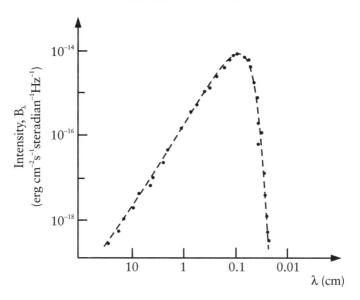

Figure 22. The observed Planck Curve of cosmic (microwave) background radiation closely obeys that of a perfect (blackbody) thermal radiator. The theoretical least-square fit (dashed) to the measured intensities $B_\lambda$ (dots) is consistent with a 2.73-K emitter; these data were obtained by the *Cosmic Background Explorer* (COBE) satellite in 1989.

lowing mathematical relation, called the Planck equation, first derived in 1900 by the German physicist Max Planck:

$$n_\lambda = 8\pi / \lambda^4 \, [e^{hc/k_B T\lambda} - 1] \, .$$

Here, among the yet undefined terms, $n_\lambda$ is the number of photons per unit volume within a narrow range of wavelengths centered on $\lambda$, h is Planck's constant ($6.63 \times 10^{-27}$ erg-s), $k_B$ is the same Boltzmann's constant used in the Introduction (equal to $1.38 \times 10^{-16}$ erg $K^{-1}$ and not the curvature constant k used for the cosmological models of Chapter One), T is the temperature (in Kelvins, K), and e is a numerical constant equal to 2.718. Analysis of the spectral data of Figure 22 enables us to solve this equation for the temperature, which yields $2.73 \pm 0.01$ K. This is the temperature characterizing the cosmic background radiation, now greatly cooled from its fiery beginnings.

In this broad review, we are mainly concerned with the density of the energy contained in radiation so that it can be compared to the energy density of matter derived above. This energy density of radiation, u, can be found by multiplying the Planck equation by the energy, $h\nu = hc\lambda^{-1}$, of each photon and integrating over all $\lambda$ to sum the contributions of all the photons to Planckian radiation,

$$u = \int_0^\infty \tfrac{8\pi hc}{\lambda^5} [e^{hc/k_B T\lambda} - 1]^{-1} d\lambda.$$

The result, which is the area under the curve of Figure 22, is[2]

$$u = a\, T^4,$$

where $a = 8\pi^5 k_B^{\,4}/15(hc)^3$ is known as the "radiation constant" and equals $7.56 \times 10^{-15}$ erg cm$^{-3}$ K$^{-4}$. For example, using the currently observed temperature of the cosmic background radiation, we find an energy density, $u = a(2.73)^4 \simeq 4 \times 10^{-13}$ erg cm$^{-3} \simeq 0.25$ eV cm$^{-3}$, corresponding to a number density of ~400 photons cm$^{-3}$. This equals less than a millionth of the power emitted by the smallest light bulb in our homes, again implying that this cosmic radiation has been greatly weakened while dispersing into an ever-increasing volume during the past 12 billion years or so. Actually, this weakening of cosmic radiation is more properly attributed to a huge Doppler effect caused by the Universe's expansion, thus shifting the frequency of the radiation clear across the electromagnetic spectrum, from its originally intense $\gamma$ rays near the time of creation to its less energetic radio waves in the current epoch. Though Doppler shifted by a factor of some 1500 times since first launched, the character and spectral shape of the Planckian radiation are unchanged by cosmic expansion. And when was that signal launched? With $z \simeq 1500$, $t \simeq 100,000$ years after the big bang.

We might wonder about other sources of electromagnetic radiation in the Universe, such as light from stars and galaxies. How do these localized radiation fields compare with the cosmic background radiation filling all space? Surprisingly, although the background radiation is extremely weak, it still contains more energy than has been emitted by all the stars and galaxies that have ever existed! This is because stars and galaxies, though very intense sources of radiation in and of themselves, occupy only a tiny fraction of all space. When their energy is averaged over the volume of the entire Universe, it falls short of the energy contained in the background radiation by at least a factor of 10. Therefore, for purposes of portraying the big picture in bulk, we can regard the cosmic background radiation as the only significant form of radiation in the Universe. Later, especially in Chapter Three, we shall reconsider those stars and galaxies on their own, more localized scales.

### Evolution of Cosmic Temperature

The arguments of the previous chapter showed how $\rho_m$ changes with time (that is, $\rho_m \simeq 10^6 \, t^{-2}$). Likewise, we now have, by manipulating the above equations of this chapter, the change of temperature with time:

$$T = 2.7(1 + z) \simeq 10^{10} \, t^{-0.5} \, ,$$

where t is expressed in seconds from the big bang. To verify this result, re-express the Friedmann-Lemaître equation in terms of $\rho_r$, the equivalent mass density *of radiation:*

$$H^2 = 8\pi G \rho_{r,0} / 3R^4 \, .$$

This is the $k = 0$, $\Omega = 1$ solution to the Friedmann-Lemaître equation, a particularly relevant approximation given that $r \ll r_0$ in the early Universe. Here the $R^4$ term owes to the radiation scaling not only as the volume ($\propto R^3$), but also by one additional factor of R, because radiation (unlike matter) is also affected linearly by the Doppler effect. Substituting for H ($= R^{-1} \, dR/dt$), integrating, solving for $\rho_r$, and using the relation $\rho_r = \rho_{r,0} \, R^{-4}$, we find

$$\rho_r = 3 / 32\pi G t^2 \, .$$

Finally, noting that the equivalent energy density of radiation, $\rho_r c^2$, equals $aT^4$, we conclude that

$$T = (\rho_r \, c^2 / a)^{0.25} \, ,$$

which, upon evaluation, equals

$$T \simeq 10^{10} \, t^{-0.5} \, ,$$

*Table 1.* Universal History of Density and Temperature

| Epoch | Time Interval | $\rho_m$ (g cm$^{-3}$) | T(K) |
|---|---|---|---|
| chaos | $< 10^{-24}$ s | $> 10^{50}$ | $> 10^{20}$ |
| hadron | $10^{-24}$–$10^{-3}$ s | $10^{30}$ | $10^{15}$ |
| lepton | $10^{-3}$–100 s | $10^{10}$ | $10^{10}$ |
| nuclear | 100 s–$10^4$ y | $10^{-8}$ | $10^7$ |
| atom | $10^4$–$10^6$ y | $10^{-19}$ | $10^4$ |
| galaxy | $10^6$–$10^9$ y | $10^{-25}$ | 100 |
| stellar | $10^9$–>$10^{10}$ y | $10^{-29}$ | 3 |

where t is again measured in seconds. Figure 23 displays the resulting $\rho_m$–T plot throughout cosmic history. Table 1 specifies the *average* values of $\rho_m$ and T for each of the seven major epochs noted at the bottom of Figure 23. To clarify, the average cosmic density would require us to pulverize mentally all the stars and other such material objects and to spread their matter evenly throughout the Universe. It is these density and temperature values, together with a sound understanding of modern physics, that grant an appreciation of key events and processes at any time in the history of the Universe.

## Descriptive Narrative of Universal History

What follows is a semi-quantitative account of the dominant activity during each cosmic epoch, based on the "numerical experiment" whose results are summarized in Figure 23 and Table 1 (where "s" denotes seconds and "y" years). All the quoted time intervals here and elsewhere are relative to our culturally invented "Earth time," which is ultimately derived from our planet's revolution about the Sun.

Most computations designed to elucidate the early Universe suggest that its physical conditions can be realistically determined to within $10^{-24}$ of the first second of existence, which is roughly the time needed for light to cross a proton. Partly because they entail considerable conjecture not subject to experimental tests and partly because they are of secondary importance to the principal thrust of this book, descriptions of events earlier than this time are relegated to a long note,[3] which can be summarized thus: Closer to the beginning of time than $10^{-24}$ second, a time interval colloquially labeled "chaos" in Table 1,[4] the above equations specify an average density greater than $10^{50}$ g cm$^{-3}$ and an average temperature greater than $10^{20}$ K. Of course, such

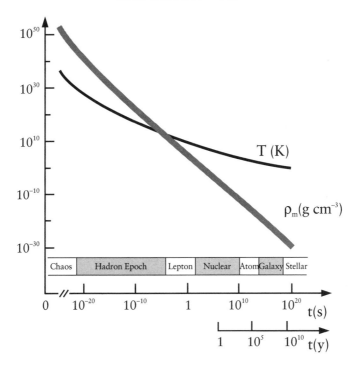

Figure 23. A log-log plot of $\rho_m$, the density of matter on average, and T, the temperature of radiation on average, over the course of all time, to date. This graph refers to nothing in particular, just everything in general. The thick width of the line drawn for $\rho_m$ represents the range of uncertainty in total matter density, whose true value depends on the amount of "dark matter"; by contrast, T is very accurately measured today, and its thin line equally accurately extrapolated back into the early Universe.

huge values are hardly comprehensible to those of us living on a planet whose familiar values include, for example, 1 and 8 g cm$^{-3}$ for the density of water and iron, respectively, and a mere 100 K for the well-known temperature changes separating ice, liquid water, and steam; even the density of an atomic nucleus is "only" $10^{12}$ g cm$^{-3}$, the temperature at the heart of a massive star "only" $10^8$ K.

The second major time interval in universal history is a bit closer to human understanding, although still characterized by severely non-terrestrial conditions. Called the hadron epoch, its name derives from the collective label given to heavy, strongly interacting elementary particles such as protons, neutrons, and mesons, which were among the most abundant types of matter at the time. Calculations suggest that well within a microsecond of its being, the Universe was filled with a whole mélange of such free, unbound, fast-moving particles that must have populated every available niche of the cosmos. Whence did they come? From radiation, pure and simple. Under the prevailing (though still somewhat surrealistic) conditions of $10^{30}$ g cm$^{-3}$ and $10^{15}$ K, these particles originated by means of a straightforward materialization—a creation—of matter from the radiative energy of the primeval fireball. Neither magic nor supernaturalism need be invoked to account for these particles, no appeal to mysticism is required; their appearance results from the well-understood and experimentally verified physical process of "pair production." In such a process, elementary building blocks of matter can naturally and spontaneously emerge from the decay of packets of high-frequency radiation—highly energetic photons—in much the opposite way that a particle and its antiparticle pair are known to yield pure energy (that is, radiation) upon collision ("pair annihilation"). For example, the average energy of a photon at the time, $2.7k_BT \simeq 0.1$ erg, or equivalently $10^{11}$ eV, would have been great enough to pair-produce a proton and an antiproton, each of which has a rest mass of nearly 1 GeV. (A proton's threshold temperature, $T \simeq 1$ GeV/$k_B$, is just over $10^{13}$ K, below which its pair production cannot occur.) Such hadrons doubtless interacted and collided with one another and with other types of fundamental particles, for the density was extremely high well within this first second of existence. Considering the inferno prevalent in the Universe at this time, such particles surely remained unbound, elementary entities; the environment was just too energized for them to have assembled into anything more structured. Accordingly, the dominant action of this epoch was the inception and then self-destruction of hadrons into high-energy radiation, thus contributing to a brilliant fireball of mostly γ rays, X rays, and (what we would now describe as) blinding light.

As the Universe continued to expand rapidly, the seethingly "hot" photons quickly dispersed, thus lowering their average energy; the creation of new hadrons from radiation became progressively less likely. Figure 23 chronicles how the cosmos, in the broadest terms, thinned and cooled. About $10^{-3}$ second after the bang, the superenergetic conditions suitable for hadron creation had ceased, allowing the lighter particles, such as electrons, neutrinos,[5] and muons, to come forth. Thus began another process of materialization that fashioned a whole new class of weakly interacting elementary particles, termed leptons. As another example of pair production, since $2.7k_BT \simeq 10^{-6}$ erg $\simeq 10^6$ eV at about this time, the average energy of a typical photon was sufficient to yield an electron and its antimatter opposite positron, each of which has a rest mass of 0.5 MeV. (An electron's threshold temperature is therefore $6 \times 10^9$ K.) By the midst of this so-called lepton epoch, the average density and temperature had fallen to some $10^{10}$ g cm$^{-3}$ and $10^{10}$ K—physical conditions in one respect greatly moderated compared to those hugely dense and intensely hot values extant a fraction of a second earlier, but in another respect still severe compared to those familiar to us on Earth. Before the first second had elapsed, the just-created leptons were mostly self-annihilating, much like the hadrons earlier, thus refueling the radiative fireball with still more high-energy photons.

Within a hundred seconds after the start of all things, the last of the newly created elementary particles had emerged from the ever-cooling fireball. Once the temperature fell below $10^9$ K, the "standard model" of big-bang cosmology maintains that the only particles remaining were protons, electrons, neutrons, and neutrinos; the Universe seems to be electrically neutral. All strange elementary particles (that can now occasionally be recreated artificially in particle accelerators) had by then disppeared. Aside from the "dark matter" whose nature and extent still elude us, no new matter has been naturally introduced since that early time.

Throughout these first few minutes of existence, equilibrium endured; the whole Universe is said to have been "thermalized." Typical "forward" and "backward" reactions and collisions (represented by the double-headed arrow), such as

$$h v \leftrightarrow \text{particle } 1 + \text{particle } 2 \, ,$$

allowed radiative photons (of energy hv) and material particles to originate and decimate at equal rates. By "equilibrium," we mean not that the radiative and material particles were equal in numbers, rather that the ratio of such particles remained constant, given the prevailing temperature and density of the cosmic medium. In fact, photons of all energies were at all times substantially more numerous than particles of any energy, the imbalance in numbers being approximately a factor of $10^9$. Such an equilibrium can be maintained only insofar as reactions of the kind described above match the pace of the rapidly changing environmental conditions. Frequent collisions among the gas and radiation must be incessant enough to keep the vast primordial sea sufficiently mixed up or disorganized, the energy everywhere spread uniformly. Eventually, several millennia thereafter, the expansion of the Universe itself would open a gap between the maximum possible randomness indicative of equilibrium and some actual, lesser randomness as the reaction rates decreased, equilibrium broke down, and order began to set in. This is true because the rate of equlibrium-maintaining reactions is more sensitive to changes in mass density ($\propto \rho_m$) than the rate of early-Universe expansion ($\propto \rho_m^{0.5}$), which tends to destroy equilibrium; as $\rho_m$ decreased with time, the former could not keep up with the latter.

Contrary to popular opinion, then, the entropy of the early Universe was already maximized *for the prevailing conditions at the time.* Extraordinary heat and extensive mixing guaranteed that thermal and chemical (really nuclear) entropies were high, yet they would have pertained only to those extreme conditions; maxima can be relative. As the Universe expanded, conditions changed dramatically; most notably the temperature and density decreased, allowing the maximum possible entropy of the Universe to grow still more (since S varies inversely with T). Furthermore, as matter eventually began to clump under the influence of gravity (recall the example of star formation discussed near the end of the Introduction), total cosmic entropy was subject to further growth, in accord with the principal implications of the second law—Nature tends irreversibly toward ever-increasing entropy. (Although gravitational entropy was low in the early Universe, structures were still non-existent because gravity was not controlling events, radiation was.) That said, however, most of the total entropy in today's Universe can be attributed to the cosmic background radiation, and if all the organized matter were to evaporate, that total entropy would change by only a fraction of 1 percent; it is in this sense that some say the universal "heat death," for the most part, occurred long ago.

Despite the onslaught of intense radiation and violent particle collisions, the seeds of construction were already beginning to emerge from the maelstrom. Throughout at least the first part of the nuclear epoch—extending from the first few minutes to no more than a few thousand years—the temperature exceeded by a wide margin the $10^7$ K minimum value needed to fuse hydrogen nuclei into helium via a neutron-proton capture scheme that first produced deuterium nuclei (a process much simpler than the proton-proton cycle known to operate in the cores of stars today). Most researchers agree that only so many deuteron particles (1 p$^+$ + 1 n$^\circ$) and thence helium nuclei (2 p$^+$ + 2 n$^\circ$) could have been synthesized in the first 15 minutes, before the average universal temperature dipped below this critical value. Despite some uncertainty about the precise values of T and ρ at the time, only about 1 helium nucleus ($^4$He) could have formed for every 12 protons ($^1$H). More $^4$He would not likely have fused from smaller particles because of a "deuterium bottleneck," owing to events in the primordial fireball happening so rapidly. There was a relative lack of *free* neutrons (n$^\circ$), which, owing to the weak interaction force, tend to break down (β decay) within 14.8 minutes, n$^\circ$ → p$^+$ + e$^-$ + $\bar{\text{v}}$, where $\bar{\text{v}}$ is an anti-neutrino. More important, $^2$He, with a nucleus containing merely 2 protons alone, cannot exist in Nature because at close range the protons' natural electromagnetic repulsion is a few percent greater than their countervailing nuclear attraction.

Fortunately for us, indeed for all life, hydrogen did not completely convert to helium. Helium is an inert element and therefore does not participate in any kind of chemistry. Also, helium needs higher temperatures than hydrogen to fuse inside stars, making most such stars short-lived, extremely hot, and intensely radiating in the ultraviolet (UV) part of the spectrum. Such short-wavelength radiation is lethal for most known life, since UV rays fracture life-related molecules, including water. Of course, if all hydrogen had converted to helium in the early Universe, then there would not have been any water ($H_2O$) either.

Numerical experiments that simulate in detail the synthesis of light atomic nuclei from the sea of protons and neutrons extant in the first minutes of cosmic evolution match remarkably well the observed abundances of normal helium ($^4$He, 9% by number), its isotope ($^3$He, $10^{-3}$%), deuterium ($^2$H, $10^{-3}$%), and also perhaps lithium ($^7$Li, $10^{-7}$%); the last of these apparently managed to leak through the above-noted bottleneck, but only in trace amounts. The Universe spent merely ~1000 seconds in the temperature range of $10^8$–$10^9$ K, appropriate for the construction of light elements. Measurements of these nucleosynthetic products are made mainly in very distant (perhaps primeval) intergalactic clouds and in the atmospheres of the oldest stars in the halo of our own Galaxy. Such good agreement between theory and observation of light-nucleus production is considered a major success of the standard big-bang model. In fact, computations of primordial nucleosynthesis, along with observations of galaxy recession and measurements of the cosmic background radiation, form the three central pillars upon which big-bang cosmology is based.

By contrast, nuclei much heavier than $^4$He could not have been appreciably produced in the early Universe. The creation of massive nuclei such as carbon ($^{12}$C), oxygen ($^{16}$O), and iron ($^{56}$Fe) requires T $> 10^8$ K; but the average universal temperature in the first minutes of the nuclear epoch was falling precipitously. Heavy-element nucleosynthesis also requires much $^4$He, since heavier nuclei are built from lighter ones; yet other bottlenecks were coming into play, such as Nature having no stable nuclei with five or eight particles. Theoretical calculations suggest that by the time sufficient helium nuclei had assembled to interact with one another to manufacture, in turn, some of their heavier counterparts, T had decreased below the just-noted threshold value needed for doubly charged helium nuclei to ignite in a fusion reaction. The heavy elements needed to await the onset of stars much later in time—small islands of hot, compact matter well suited to reverse the cooling and thinning conditions of the early Universe.

Of central import for us in this book—and an argument that we shall quantitatively examine in the next section—the violent conditions of the primordial Universe guaranteed that the energy housed in radiation greatly exceeded that effectively contained in matter. Even the nuclei just noted were an insignificant "contaminant" at the time. Not only did the photons of radiation far outnumber any particles of matter, most of the energy in the early Universe took the form of radiation, not matter. As soon as the elementary particles of matter tried to assemble into anything much more substantive, high-frequency radiation (aided by violent particle collisions) destroyed them, thus precluding the existence of even the simplest structures we now call "atoms," let alone any stars or galaxies. For this reason, the first four epochs of Table 1 and Figure 23 collectively comprise the Radiation Era.[6] Whatever matter managed to exist in the Radiation Era did so as a relatively thin microscopic precipitate suspended in a macroscopic, glowing "fog" of dense, brilliant radiation.

As time elapsed, change continued. The fifth major phase of Table 1—the atom epoch—extends in time from about 10,000 years to perhaps as much as a million years after the bang. Midway through this period, the average density had decreased to a value of nearly $10^{-20}$ g cm$^{-3}$, while the average temperature had fallen to some $10^4$ K—values much like those in the atmospheres of stars today. A principal feature of the atom epoch was the steady waning of the original fireball. The Universe by this time had expanded considerably, the annihilation of hadrons and leptons had ended, and primordial nucleosynthesis had all but ceased. Even as the fireball faltered, though, a dramatic change began.

For the first hundred centuries, radiation had reigned supreme over matter. All space was flooded with photons, especially light, X rays, and γ rays, ensuring a non-structured, undifferentiated, virtually informationless, and highly uniform blob of plasma. We say that matter and radiation were intimately coupled to each other—thermalized and equilibrated. Particle symmetry reigned supreme. As the universal expansion paralleled the march of time, however, the energy housed in radiation decreased faster than the energy equivalently contained in matter. This imbalance ultimately caused that opaque, luminous fog of energetic photons to dim and then to lift, thus diminishing the early dominance of radiation. Matter and radiation thereafter began to decouple, their equilibrium unraveling and symmetry breaking, as an evolutionary event of first magnitude set in.

Sometime between the first few millennia and a million years after the bang—an exact moment cannot be established largely because of the statistical nature of the gradual change from an ionized state to a neutral state—the charged elementary particles of matter clustered into more structured atoms. Laboratory studies have shown that at $T \simeq$ 4000 K, collisions are sufficiently violent and frequent to shatter lightweight atoms; at lower temperatures, collisions are insufficient to do so.[7] And according to the above formulae, $T = 4000$ at $t \simeq 100,000$ years (corresponding to $z \simeq 1500$); that's a short time interval relative to the current age of the Universe, yet 100 millennia in absolute time. Owing to their *charged* nature, the particles' own electromagnetic forces caused leptons and hadrons to combine, sporadically at first and then more frequently, since the lower-frequency radiation (and weakened particle collisions) could no longer split the atoms as quickly as they formed. Matter, as it neutralized, finally gained some leverage in a Universe previously ruled by pure energy; neutral hydrogen in particular interacts very weakly with blackbody radiation at this temperature. In a sense, matter had managed to overthrow the cosmic fireball and emerge as the dominant constituent of the Universe.

This phase transition from the ionized to the neutral state—an evolution from radiation-dominance to matter-dominance—signifies as fundamental a change as has ever occurred in all of history. In fact, matter's rise to dominance is the first of two preeminent changes throughout cosmic evolution. (The other great change—the onset of technologically sentient life forms—will be encountered in Chapter

Three). To denote this major turn of events, the last three epochs of Table 1 are collectively termed the Matter Era.

Once the Matter Era was fully established, atoms were literally everywhere. The disruptive influence of radiation and the disorganizing effects of collisions had grown so weak that they could no longer prevent the attachment of the leptons and hadrons that had survived annihilation. Accordingly, hydrogen atoms were among the first neutral elements to form, requiring only that a then slower-moving negatively charged electron be electromagnetically linked to a positively charged proton. Copious quantities of atomic hydrogen were thereby made in the early Universe, and it is for this reason that we regard hydrogen as the common ancestor of all material things.[8]

By the end of the atom epoch, the Universe had evolved dramatically. The spectacularly bright fireball associated with the immediate aftermath of creation had waned; the previously opaque cosmos had become transparent. Some say that the Universe had entered the real "dark ages," when no sources of light whatsoever existed anywhere—no stars, galaxies, or anything else to illuminate the gloomy blackness. The physical conditions of temperature and density that guide all changes in the Universe had themselves undergone extraordinary change. And matter had wrested firm control from radiation, thus heralding a whole new era.

Thereafter, major events in the Universe were less frequent. Change continued unabated, but at a more relaxed pace. For once the Universe had cooled enough to allow the formation of atoms (albeit thinly dispersed ones), subsequent events necessarily occurred more slowly. Eventually, though surely not before about a million years after the bang, gravitational instabilities and statistical fluctuations caused some of the matter to assemble into vast clumps. Observed inhomogeneities in the cosmic background radiation (at the level of less than one part in $10^5$) imply that the earliest gravitationally bound collections of matter would have been of order $10^6$ $M_\odot$; whether these massed aggregates formed the first generation of star clusters, presumably now long gone, or whether they became the cores of today's most distant quasars is unknown.

During the sixth, or galaxy, epoch, galaxies and galaxy clusters began forming, probably by hierarchical means that built large structures from smaller entities—a bottom-up approach to structure—but the details are not yet fully understood.[9] What we do know with reasonable certainty is that the quasars and remotest galaxies must have emerged in the earliest parts of this period. Indeed, all the galaxies must have originated long ago, for observations imply that none has formed within the past 10 billion years or so. Midway through the galaxy epoch, according to our computations in this chapter, the average (matter) density of the Universe had decreased by about another factor of a million, reaching some $10^{-25}$ g cm$^{-3}$; the average (radiation) temperature had also declined to a relatively cool 100 K. The entire Universe was growing ever thinner, colder, and darker.

Events had slowed up considerably by this time, some billion years after the bang. Although the early Universe was characterized by rapid change, especially in the first few minutes of the Radiation Era, the later Universe changed more sedately. But it changed nontheless; natural history was, and is, still being written.

Finally, in our sequence of seven temporal intervals stretching across all time, we have the stellar epoch—the one now engulfing us in space and time. Not that stars were unable to have existed in the earlier galaxy epoch; a first generation of massive stars might well have emerged and then quickly supernovaed. But as the name implies, the dominant action of this current period is the origin of stars and the synthesis of heavy nuclei inside stars—objects intermediate in size between atoms and galaxies. Of this we are certain, for research during the past two decades has provided direct observational evidence that new stars are actually now forming from galactic gas and dust; galaxies themselves are apparently not forming in the current epoch, but stars within them quite definitely are. At the present time, the average density of the Universe is more than another thousand times thinner than in the previous galaxy epoch, now approximately $10^{-29}$ g cm$^{-3}$. As derived earlier, this is the critical value above which the Universe will eventually contract back to a point much like that from which it began ($k = +1$, $\Omega > 0$), and below which the Universe will expand forevermore ($k = -1$, $\Omega < 0$). Subtle observational tests, now under way, have not yet specified the precise density, nor therefore do we know the ultimate fate of the Universe (consult again note 5 in this chapter). And as for the current temperature of everything in the cosmos, galaxies, stars, and empty space alike average a few Kelvin—the cold relic of the awesomely hot fireball prevalent eons ago, the fossilized grandeur of an ancient and glorious era. It's neither a figment of our imaginations nor an untested prediction of our Universe models; as noted earlier in this chapter, the 2.7-K cosmic background radiation has been clearly detected with radio telescopes and orbiting satellites. It is one of the strongest features of modern big-bang cosmology.

Mentioned here only for completeness, interesting by-products of the stellar epoch are the clusterings of matter into planets, life, and intelligence, much of which are treated en masse in Chapter Three.

## Comparison of Energy Densities

Let us now make the final computation of this chapter, arguably the most important one thus far deduced. It will serve to justify some of the key statements made in this chapter while describing the cooling and thinning of the early Universe. Here we compare the "equivalent" energy density of matter, $\rho_m c^2$ (derived in the previous chapter), with the "pure" energy density of radiation, $aT^4$. Note, for example, that at the present time $\rho_{m,0} c^2 \simeq 2 \times 10^{-9}$ erg cm$^{-3}$ (which is nearly ten times less dense than the theoretical $k = 0$, $\Omega = 1$ case, since not enough matter has been observationally found to justify the $k = 0$ solution); whereas $aT_0^4 \simeq 4 \times 10^{-13}$ erg cm$^{-3}$. In other words, during the current epoch, $\rho_{m,0} c^2 > aT_0^4$ by several orders of magnitude, proving that matter is now in firm control (gravitationally) of cosmic changes, despite the Universe still being flooded today with radiation. (In fact, the number of photons of radiation far exceeds the number of particles of matter even some 12 billion years after creation, yet the energy per photon has been greatly reduced.) If enough dark matter does exist to comply with the $k = 0$, $\Omega = 1$ model, then the inequality, $\rho_{m,0} c^2 > aT_0^4$, is strengthened by an additional factor of ten, and the conclusion that matter now dominates radiation is secure.

A key issue here is that at other times in the history of the Universe these two energy densities were not always in such discord. Not only do these densities themselves change as the Universe evolves, but it so happens that their ratio also changes with time. To be sure, there must have been a time in the past when $\rho_m c^2 = aT^4$, and an even earlier time when $\rho_m c^2 < aT^4$.

To confirm this, first recall from Chapter One that the matter density scales inversely as the volume, so that the equivalent energy density varies in the same manner,

$$\rho_m c^2 = \rho_{m,0} c^2 R^{-3} .$$

By contrast, as also noted earlier, the energy density housed in radiation scales as an additional factor of R because radiation (unlike matter) is also linearly affected by the Doppler effect. With the expansion of space, the wavelength of every photon grows, much as though the wave were printed on the surface of an expanding balloon; this is the cause of the cosmological red shift, which further reduces each photon's energy density over and above its normal dilution in an increasing volume. Hence, we have

$$aT^4 = aT_0{}^4 R^{-4} .$$

For these relations, as in the previous section, the subscript "0" represents variables at the current time; those variables without a subscript represent values at any time.

We can now ask: When were the two energy densities precisely equal, namely, $\rho_m c^2 = aT^4$? Expressing both of them in terms of their current values,

$$\rho_{m,0} c^2 / aT_0{}^4 = R^{-1} ,$$

or, from the previous section, since $R^{-1} = (1 + z)$, we find upon evaluating for the (above-computed) current values of $\rho_{m,0} c^2$ and $aT_0{}^4$ a value of $z \simeq 5000$. And from the equation earlier in the chapter that connects Doppler shift and time, $2.7(1 + z) \simeq 10^{10} t^{-0.5}$, we calculate that this z corresponds to $t \simeq 10,000$ years. Thus, less than a million years after the big bang, the two energy densities were equal; at earlier times, $\rho_m c^2 < aT^4$. (Again, if dark matter exists in quantities sufficient to make $\Omega \simeq 1$, then the two energy densities would have equaled earlier, namely, at $z \simeq 25,000$ and $t \simeq 1,000$ years. The essential qualitative arguments remain the same whether or not dark matter is present; quantitatively, the effect of dark matter is to slide back or forth some of the key events along the temporal axis.)

Figure 24 is of vital significance in our discussion, indeed for the central theme of this book. In part, it reproduces the temporal behavior of $\rho_m c^2$, as drawn in Figure 20. Superposed on it is the temporal change of $aT^4$ computed from the equations of this chapter. Notice how the two values intersect at $t \simeq 10,000$ years, the epoch when the energy distributed between matter and radiation was last equilibrated. Relative to this crucially important turning point in cosmic history, pure energy (radiation) dominated in earlier epochs while equivalent energy (matter) has ruled in later epochs.

This crossover in energy densities represents probably the most important change in all of cosmic history. The event, $\rho_m c^2 = aT^4$, separates

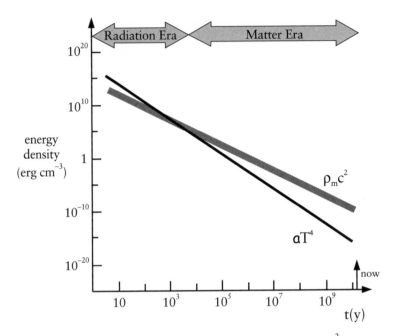

Figure 24. The temporal behavior of both matter energy density, $\rho_m c^2$, (taken from Figure 20) and radiation energy density, $aT^4$. The two curves intersect at t ≃ 10,000 years, at which time the atom epoch began, the Universe started to change dramatically, and the Radiation Era transformed into the Matter Era. The plot of $aT^4$ is precise because T is known to high accuracy; that for $\rho_m c^2$ is "fuzzy" because $\rho_m$ is uncertain (owing to the unresolved issue of "dark" matter), and so the top of the curve for $\rho_m c^2$ corresponds to $\Omega = 1$, the bottom to $\Omega = 0.1$.

the Radiation Era from the Matter Era, and designates that time at which the Universe gradually began to become transparent. Some 10,000 years after the start of all things, both thermal equilibrium and particle symmetry began to founder, causing the radiative fireball and the matter gas to decouple. It was as though a bright, opaque fog (much like the gas discharge inside a neon sign) had begun to lift. Photons, previously scattered innumerable times by subatomic material particles (especially free electrons) of the expanding, hot, opaque plasma in the Radiation Era, were no longer so affected once the electrons became bound into atoms in the Matter Era. (See again note 8 in this chapter.) This most dramatic change was practically over by about 100,000 years, when the last throes of the early plasma state had mostly transformed into neutral matter. A uniform, featureless state was overthrown by one in which order and complexity were thereafter possible.

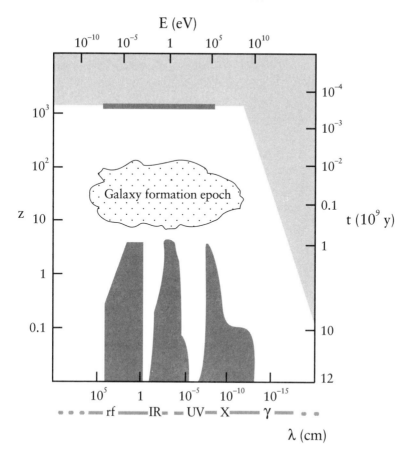

Figure 25. This plot of "all possible electromagnetic knowledge" depicts the extent to which the Universe can be currently explored by electromagnetic means. Red shift, z, and time, t, from the big bang are graphed vertically, whereas photon energy, E, and wavelength, λ, are plotted horizontally. Note that the average energy of a blackbody photon equals $2.7k_BT$ expressed in ergs, and that 1 electron volt, eV, equals $1.6 \times 10^{-12}$ erg. The heavily shaded areas, including the horizontal bar at $z \simeq 1500$, indicative of the cosmic background radiation, represent the domains of current knowledge. The lightly shaded area is "elsewhere," which cannot be directly observed, partly (at great z) because of electron scattering of primordial photons and partly (at high E) because of the opacity of nearby gamma-rays due to pair production. The epoch of galaxy formation is represented by the irregular blob at the center for which $z > 10$; the origins of galaxies represent perhaps the greatest missing link in all of cosmic evolution, largely because humankind has yet to build a machine capable of imaging realms so far away and so long ago.

The 2.7-K radiation reaching Earth today is a relic of this signal phase transition, sometimes called the "surface of last scattering." The microwave radiation now captured by radio telescopes and orbiting satellites has streamed unimpeded (save for being greatly red-shifted) across space and time for most of the age of the Universe, granting us a "view" of this grandest of all evolutionary events that occurred long, long ago.

## Summary of Chapter Two

Current studies of material objects have been unable to probe any closer to creation than about a billion years (corresponding to about 90 percent of the "look-back time"). By contrast, analyses of the cosmic background radiation push back our knowledge to some hundred thousand years after the origin—a mere hundred millennia. This means that observations of the background radiation help characterize directly, if broadly, the most recent 99.999 percent of cosmic history. To probe closer to the big bang itself, we are forced to rely on theory, albeit *scientific* theory in which observational and experimental results play a prominent yet indirect role.

While the recession of the galaxies implies that at some past time the Universe began as a compact, hot, primal blob, the cosmic background radiation virtually proves it. Together with the remarkable agreement between theoretical and observational studies of primordial nucleosynthesis, these two key observations—of material objects and of radiative energy—strongly support an evolutionary Universe. Figure 25 expansively sketches the status of radiation and matter thus far considered. These two vital pieces of cosmic inventory virtually assure the validity of some version of big-bang cosmology. It is within the context of this "standard model" that we can begin to understand the emergence, growth, and evolution of intricately complex structures, a premier example of which is life.

# CHAPTER THREE

# LIFE

O f all the known clumps of matter in the Universe, life forms, especially those enjoying membership in advanced technological civilizations, arguably comprise the most fascinating complexities of all. What is more, technologically competent life differs dramatically from lower forms of life and from other types of matter scattered throughout the Universe. This is hardly an anthropocentric statement; after more than 10 billion years of cosmic evolution, the dominant species on planet Earth—the human being—has learned to tinker not only with matter and energy but also with evolution. Whereas previously the gene (strands of DNA) and the environment (whether physical, biological, or cultural) governed evolution, twenty-first-century Earthlings are rather suddenly gaining control of aspects of both these agents of change. We are now tampering with matter, diminishing the resources of our planet while constructing the trappings of utility and comfort. And we now stand on the verge of manipulating life itself, potentially altering the genetic makeup of human beings. The physicist unleashes the forces of Nature; the biologist experiments with the structure of genes; the psychologist influences behavior with drugs. We are, quite literally, forcing a change in the way things change.

But what is life? Like time, life is obvious to discern yet elusive to define. Although most biologists generally skirt the issue, we suggest that our very essence can be defined as follows:[1] Life is an open, coherent, spacetime structure maintained far from thermodynamic equilibrium by a flow of energy through it—a carbon-based system operating in a water-based medium, with higher forms metabolizing oxygen. Although the second part of this definition pertains to the living state as we know it, the first part could well apply to a galaxy, star, or planet (much as I do in fact define those terms in the Glossary). And that is a crux of our argument: Life likely differs from the rest of clumped matter only in degree, not in kind. We admit no vitalism, no special life force that would set animate beings manifestly apart from all other forms of inanimate complexity. Nor do we include the property of replication in our definition of life (Morrison 1964); some organisms do not replicate, including mules, sterile men, and individual humans (two are needed), whereas many non-living systems can be claimed to do so, such as flames burning, crystals growing, and even stars spawning other stars. All these inanimate systems are as reasonably conceivable replicators as any collection of bugs in a petri dish. Nor, alas, do we embrace in our definition of life the notion of autopoiesis (Maturana and Varela 1988)—namely, an innate "process" that, in addition to structure and function, acts as a third key operational aspect of life, yet one that diminishes the role of interactions with the environment. For us, process is essentially synonymous with the flow of energy, a physical effect that displays decidedly vital interactions with the environment.

That said, the degree to which living systems differ from non-living systems can cause thresholds to be crossed, as in the case of sentient life, like us. The emergence of technologically intelligent life, on Earth and perhaps elsewhere, heralds a whole new era: a Life Era. Why? Because technology, despite all its pitfalls, enables life to begin to control matter, much as matter evolved to control radiation more than 10 billion years ago. Accordingly, matter is now losing its total dominance, at least at those isolated residences of technological society, such as on planet Earth. To use a cliché, life is now taking matter into its own hands—a clear case of mind over matter, without any Cartesian separation asserted or implied. Such a moderate reductionist viewpoint, materialistic yet not deterministic, embodies holism as well, for here we postulate a continuous spectrum of complexity all the way up and down the line, from amorphous and unadorned protogalaxies to socially stratified cultures of extraordinary order.

A central question before us is this: How did the neural network within human beings acquire the complexity needed to fashion societies, weapons, cathedrals, philosophies, and the like? To appreciate the quintessence of life's historical development, especially of life's evolving dominance, we resume our study of the cosmic environment, broadly considered. And here we return to some of the thermodynamic issues raised earlier.

## Diverging Temperatures

As noted in the previous chapter, prior to decoupling, when matter and radiation were still well mixed in the Radiation Era, the average temperature of the Universe varied inversely as the square root of the time, t. A single temperature at any time is sufficient to describe the early thermal history of the Universe, for the overgreat densities then produced so many collisions as to guarantee an equilibrium. During this early era, disorder reigned supreme and the traditional form of entropy was maximized, but this is thermal and chemical entropy; gravitational entropy, as noted toward the end of the Introduction, was not at all high in the Radiation Era. The absence of any temperature gradients between radiation and matter dictated virtually zero information, or nearly nil macroscopic order, in the early Universe. All was uniform, equilibrated, and boring.

Once the Matter Era began, however, matter became atomic, the gas-energy equilibrium was destroyed, and a single temperature is no longer enough to specify the bulk evolution of the cosmos. Two temperatures are needed: one to describe radiation and another to describe matter. It so happens that the derivation in Chapter Two is valid throughout all time for radiation, which is effectively an expanding relativistic gas consisting purely of photons. Recalling that the observed T is red shifted in precisely the same manner as is the frequency of an individual photon, and since $u \propto T^4$ and also $u \propto R^{-4}$, we now rewrite

$$T_r \propto R^{-1},$$

or

$$T_r \simeq 10^{10} t^{-0.5},$$

with $T_r$ being the average "temperature of radiation" at any time, t, expressed in seconds.

By contrast, once (re)combined in the neutral state, matter cooled much faster and (at least for neutral hydrogen and helium below 4000 K) obeyed the relation for a perfect, non-relativistic gas. For such an ideal gas, the equation of state, $PV = R_g T_m$; also for adiabatic ($\delta Q = 0$)

expansion, $PV^{5/3}$ = constant. Therefore, the average "temperature of matter," $T_m \propto V^{-2/3} \propto (R^3)^{-2/3}$, so that for t > 100,000 years after the big bang,

$$T_m \propto R^{-2},$$

or,

$$T_m \simeq 6 \times 10^{16} \, t^{-1}.$$

We do not mean to imply a universal thermal distribution for neutral intergalactic matter akin to the detected background radiation for photons; as noted two paragraphs hence, astronomers are uncertain of the state of this loose dark matter. While $T_r$ would have fallen by a factor of ~1500 between decoupling and the present epoch—from 4000 K to 2.7 K—$T_m$ would have theoretically fallen far more, in fact by as much as a factor of a million. If so, $T_m$ for intergalactic matter would now be only about a thousandth of a degree Kelvin, again an average temperature of all matter in bulk.

Figure 26(a) illustrates these diverging thermal histories as the Matter Era evolves. Much as for the steady decrease of T and $\rho_m$ derived earlier (Figure 23), these temperature profiles refer to nothing in particular, rather everything in general—in this case, the widespread environmental conditions in the Universe after decoupling. As noted in the historical narrative of the previous chapter, the reason for the rapid cooling of the matter is that the exchange of energy between the radiation field (photons) and the gas particles (mainly H and He atoms) failed to keep pace with the rate of general expansion of the particles away from one another. The equilibrating reaction rates ($\propto \rho_m$) fell below the cosmic expansion rate ($\propto \rho_m^{0.5}$) and non-equilibrium states froze in. Thus, once the grand symmetry was broken, the thermal decline of matter exceeded, all the while its energy density decline lagged, the corresponding values of the radiation field—a curious result, yet one that enabled matter, in the main, to "build things."

Alas, unclustered intergalactic matter might have suffered an alternative fate. Some theoretical studies (Haiman and Loeb 1998) imply that, within a billion years after decoupling, neutral intergalactic matter might have been reheated, and perhaps even reionized to the plasma state, by intense energy sources such as the primeval quasars blazing forth in their youth as well as ultraviolet radiation from the first gener-

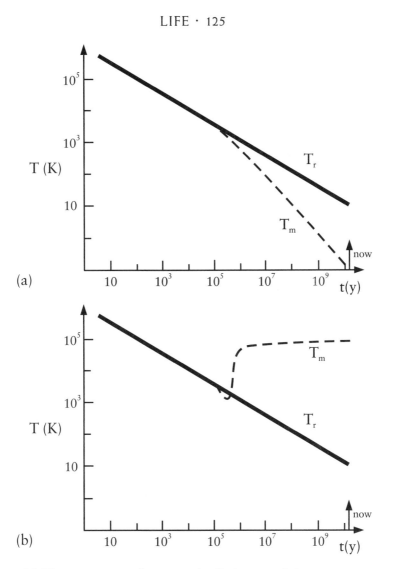

Figure 26. The temperatures of matter and radiation went their separate ways once these quantities became fully decoupled at t ≃ 100,000 years. Since that time, the Universe has been in a non-equilibrium state. Frame (a) is the case of $T_m$, the average temperature of matter, decreasing monotonically, whereas frame (b) shows the possible dramatic reheating of loose intergalactic gas within a billion years after decoupling. Either way, whether the loose gas steadily cooled or rapidly heated, a definite temperature gradient developed between radiation and matter in the expanding Universe, a clear manifestation of non-equilibrium conditions prevailing throughout.

ation of massive star clusters now mostly gone. The circumstantial evidence favoring such a grand reheating derives mainly from the observation that highly red-shifted absorption lines toward distant quasars are relatively metal-rich (though only about a hundredth that in the Sun), and these heavy elements were likely produced inside an early population of massive stars. However, such reheating events, subsequent to recombination, would have likely produced distortions superposed on the cosmic microwave background—distortions that seem to be absent; furthermore, no such stars with zero heavy elements have ever been found. In any case, Figure 26(b) shows how the run of $T_m$ might have played out over time if this reheating did occur, the overall average temperature of matter rising again to some $10^4$–$10^5$ K, comparable to nebular gas engulfing today's star-forming regions. Despite its great temperature, this loose intergalactic gas that never managed to accumulate into well-formed structures would today be invisible, owing to its extremely low density well beyond the galaxies and galaxy clusters—density on the order of at least a million times thinner than the already vacuum-like gas of interstellar space within the galaxies, and less than that needed (in the form of dark matter) to close the Universe.[2]

Regardless of whether $T_m$ became, and is now, greater or less than $T_r$, the difference between these two temperature fields would have grown as the cosmos, on average, departed gradually from its original equilibrium state. The establishment of a cosmic temperature gradient is the essential point. Such a thermal gradient is the patent signature of a heat engine, and it is this ever-widening gradient that has rendered environmental conditions suitable for the growth of complexity. The result is a grand flow of energy between the two differentiated fields, and with it a concomitant availability of energy (for use in work) over and above that extant in the early, equilibrated Universe. Hence, the Matter Era has become increasingly unequilibrated over the course of time; the expansion of the Universe guarantees it. Such non-equilibrium states are suitable, indeed apparently necessary, for the emergence of structure, form, or organization—of order! Thus we reason that *cosmic expansion itself is the prime mover for the construction of a hierarchy of complex entities throughout the Universe.* In what follows, we briefly quantify in the broadest terms the evolutionary development of order in the Matter Era from the disorder of the Radiation Era; the result is the Life Era.

## More on Information, Entropy, and Negentropy

As discussed earlier, the extent to which a system is ordered can be estimated by appealing to information theory; the more ordered a system, the greater the information content (or negentropy) it possesses. To be more specific, imagine a system that has a large number, $B_i$, of possible initial structural arrangements. Suppose, further, that after some information is finally received, the number of possible states of the system is reduced to $B_f$, so that $1 \leq B_f \leq B_i$; this is true because additional information about the state of the system is at hand, in particular information about its structure. Quantitatively, these statements can be expressed in terms of the net information transmitted, $\delta I$, which equals the difference between the initial and final informational states of the system:

$$\delta I = I_f - I_i = K \log (B_i / B_f) .$$

Here we have made use of the Shannon-Weaver formula stipulating that information depends inversely on the logarithm of the probability distribution of states within a system; when $B_f$ is small (after some information has been received), $\delta I$ is large, and conversely. Note that this equation increases as $B_i$ increases and $B_f$ decreases, that it maximizes for a given $B_i$ when $B_f = 1$ and vanishes when $B_i = B_f$, and that it makes information received in independent cases additive. This accords well with human intuition and empirical findings.

The above relation for information exchange has an uncanny resemblance to that describing a change in entropy, $S$, for a system having different numbers of microscopic states, $W$, before (i) and after (f) the change, to wit,

$$\delta S = S_f - S_i = k_B \ln (W_f / W_i) .$$

Manipulating these two equations, substituting $S_i k_B^{-1}$ for $\ln B_i$ and $S_f k_B^{-1}$ for $\ln B_f$, and realizing that the factor of 2.3 difference between the natural and base-10 logarithms can be absorbed within the new constant $(K k_B^{-1})$ term, we find an explicit connection between entropy and the net information exchanged,

$$\delta I = K k_B^{-1} (S_i - S_f) .$$

And since $S_i = S_{max}$, because the initially large number of receivers (microstates) relates to the maximum entropy, the information gain of a system can be expressed as the difference between the system's maximum possible entropy and its actual entropy at any given time—namely,

$$I = K\, k_B^{-1}\, (S_{max} - S)\,.$$

This is the origin of the well-known argument that order, information gain ($+\delta I$), and negentropy ($-\delta S$) are intimately related. This is also consistent with the definition of order as an absence of disorder, for the amount of order present in any system can be measured by the size of the gap by which the actual randomness differs from the maximum possible randomness (Frautschi 1982; Brooks and Wiley 1988; Layzer 1988).

Now we can apply these ideas to the central query put forth earlier in the Prologue. What is the source of order, form, and structure characterizing all material things? Recall the mathematical expression for the second law of thermodynamics, $\delta S = \delta Q/T$, meaning that for a given amount of heat transferred (between two systems), $S_{max}$ will occur for minimum $T$; for an evolving Universe, this is the system of matter characterized by $T_m$, just to take as an example the case in which the young Universe did not reheat during the early galaxy epoch. By contrast, a smaller, actual $S$ will pertain to the system of radiation defined by $T_r$, which after the decoupling (or recombination) phase around $10^5$ years, always exceeds $T_m$. Thus, by virtue of this finite temperature difference between matter and radiation, as well as the associated irreversible flow of energy from the radiation field to material objects, we obtain a quantitative measure of the potential for information growth throughout cosmic history:

$$I = K\, k_B^{-1}\, (T_m^{-1} - T_r^{-1})\,.$$

As asserted earlier, the necessary (though not necessarily sufficient) condition for the growth of information is guaranteed by the very expansion of the Universe. The Universe self-generates a thermal gradient, and increasingly so with time, suggestive of an ever-powerful heat engine were it not for its mechanistic inference. To be sure, we must emphasize throughout the statistical nature of all these processes, meaning that the growth of order is not a foregone conclusion, nor is the Universe a machine. As already noted, thermodynamics tells us if events can occur, not whether they actually will occur. Likewise, this is *potential* information, realized only should Nature take advantage of the newly established conditions for the development of systems.

With the temporal functions for $T_m$ and $T_r$ known, the evolution of I can be traced in much the same way that the change of $\rho_m c^2$ and $aT^4$ were graphed earlier. Figure 27 sketches such an I–t diagram, as well as the evolution of negentropy from which it derives. Notice how the information content was essentially zero in the early Universe when matter and radiation were equilibrated; the entropy change then was also zero, even as the Universe continued to expand. This is as it should be, for the early Universe would have experienced adiabatic expansion in the absence of any significant radiation sources; with $\delta Q = 0$, then $\delta S = 0$. After decoupling, however, the potential for the growth of information rose steadily as $T_m$ and $T_r$ diverged when the matter and radiation steadily departed from equilibrium; I has become substantial in recent times. Thus we confirm the seemingly paradoxical yet wholly significant result for the scenario of cosmic evolution: In an expanding Universe, both the disorder, S, and the order, I, can increase simultaneously—a fundamental duality, strange but true.

Early on, when the primordial nuclei began falling out of equilibrium, their reactions still generated entropy increases (mainly by creating new photons and neutrinos, as would later occur in the cores of stars), but these increases were insufficient to reestablish equilibrium. Accordingly, the gap between the actual S and the maximum possible S that would have been achieved had equilibrium been restored increased even more, thereby building up free energy, to be released much later when stars formed. Here, then, we gain insight into the origin of free energy; it is not new energy as such, rather newly rearranged energy thereafter available for use in the course of evolution. The cosmologist David Layzer (1975) perhaps put it best:

Suppose that at some early moment local thermodynamic equilibrium prevailed in the universe. The entropy of any region would then be as large as possible for the prevailing values of the mean temperature and density. As the universe expanded from that hypothetical state the local values of the mean density and temperature would change, and so would the entropy of the region. For the entropy to remain at its maximum value (and thus for equilibrium to be maintained) the distribution of energies allotted to matter and to radiation must change, and so must the concentrations of the various kinds of particles. The physical processes that

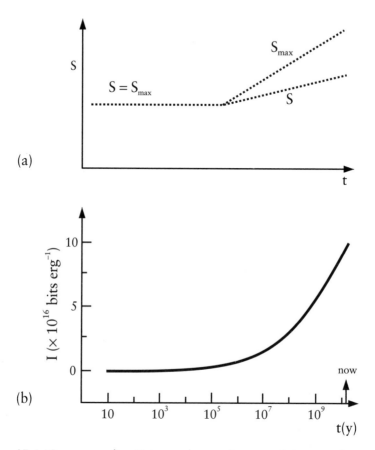

Figure 27. (a) In an expanding Universe, the actual entropy, S, increases less rapidly than the maximum possible entropy, $S_{max}$, once the symmetry of equilibrium was broken. By contrast, in the early, equilibrated Universe, $S = S_{max}$ for the prevailing conditions. (b) The potential for the growth of information content, I—related to the difference between the two values of entropy—has increased ever since decoupling. Accordingly, the expansion of the Universe can be judged as the ultimate source of order, form and structure, promoting the evolution of everything in the cosmos.

mediate these changes proceed at finite rates; if these "equilibrium" rates are all much greater than the rate of cosmic expansion, approximate local thermodyanamic equilibrium will be maintained; if they are not, the expansion will give rise to significant local departures from equilibrium. These departures represent macroscopic information; the quantity of macroscopic information generated by the expansion is the difference between the actual value of the entropy and the theoretical maximum entropy at the mean temperature and density.

The very expansion of the Universe, then, provides the environmental conditions needed to drive order from chaos; the process of cosmic evolution itself generates information. How that order became manifest specifically in the form of galaxies, stars, planets, and life has not yet been deciphered in detail; that is the subject of many specialized areas of current research. We can nonetheless identify the essence of the development of natural macroscopic systems—ordered physical, biological, and cultural structures able to assimilate and maintain information by means of local reductions in entropy—in a Universe that was previously completely unstructured.

Furthermore, because the two temperatures characterizing the Matter Era diverge, the growing departure from thermodynamic equilibrium allows the cosmos to produce *increasing* amounts of net entropy *and* macroscopic information. We thereby have a means to appreciate in the main, if perhaps not yet understand the particulars, the observed rise in complexity throughout the eons of cosmic evolution. To be sure, we have not overlooked the similarity of the culminating graph in Figure 27(b) with that of our initial impression in Figure 3, a pair of sketches that clearly address this book's core objective: To explain the growth of order, form, and structure among all materiality, yet to do so by identifying specific, quantitative ways of generating substantial amounts of organization, negentropy, and information characterizing an immense number of complexities as intricate as the rings of planet Saturn or the insects of a tropical forest, let alone the labyrinthine architecture of the human brain.

We end this section by reiterating a central goal, to quantify the general conditions for the growth of complexity on all scales. The basic idea has been posed, albeit qualitatively, over the ages, even one that presupposes the origin and existence of life in a universal setting. Witness the nineteenth-century French physician Claude Bernard (1885, as quoted in Lotka 1924): "It is not by struggling against cosmic conditions that the organism develops and maintains its place; on the contrary, it is by an adaptation to, and agreement with, these conditions. So, the living being does not form an exception to the great natural harmony which makes things adapt themselves to one another: it breaks no concord; it is neither in contradiction to nor struggling against general cosmic forces; far from that, it forms a member of the universal concert of things, and the life of the animal, for example, is only a fragment of the total life of the universe."

## Free Energy Rate Density

We are converging on an answer to the foremost question at hand: Have the many varied real structures known to exist in the Universe displayed the sort of regular increase in order and complexity during the course of time as suggested by human intuition (Figure 3) and implied by theoretical analysis (Figure 27)? The answer is yes, and more. Yet how shall we best characterize that order? What single common term might be used to quantify order on all spatial and temporal scales? In appealing to the real world surrounding us, it is perhaps best to avoid use of the term information. When examined on a system-by-system basis, information content can be a slippery concept full of dubious semantics, ambivalent connotations, and subjective interpretations (Wicken 1987; Brooks and Wiley 1988; Marijuan, Conrad, et al. 1996). Especially tricky and controversial is *meaningful* information, the value of information. What is a meaningful message or a meaningfully ordered structure, as opposed to information in a system that is of no particular use to that system, or that which is unrelated to the behavior of the system? For example, an entire DNA strand made of only one type of nucleotide base does contain raw information, but hardly any programmatic information of the type needed to manufacture proteins. Furthermore, as some researchers claim, perhaps information does need to be interpreted for it to exist, in which case it would presumably be associated only with biological and social systems; yet the telephone company doesn't care about the content of its transmitted messages, charging only by the length of the information string—a wise and appropriate way to bill its customers if it wishes to stay in business. The conceptual idea of information has been useful, qualitatively and heuristically, as an aid to appreciate the growth of order and structure in the Universe, but this term is too vague and subjective to use in quantifying a specific, empirical metric describing a whole range of real-world systems. Nor do we feel any compelling need to treat information as a third, fundamental ingredient of the Universe—after matter and energy—for, to us, information basically *is* a form of energy, whether flowing, stored, or unrealized.

Likewise, the term negative entropy (or "negentropy") is well-nigh impossible to measure, alas even to define adequately in regard to non-equilibrium states. Even its champion, Erwin Schrödinger, noted his uneasiness with the term in the reprinting of his classic essay, *What is Life?* Entropy, whether positive or negative, holds true only for static, equilibrium conditions. Approximations are possible for changing states, but only in local regions where temperature, pressure, and other such quantities change so slowly as to be measurable in small space-wise (and timewise) neighborhoods. These are so-called LTE, or "Local Thermodynamic Equilibrium," mini-states wherein a system's macroscopic changes are assumed to occur much more slowly than its microscopic changes. Hence, entropy measures (even if they could be made) involve a technicality based on an assumption, in fact one that might not always prevail for rapidly changing evolutionary events. (See Corning and Kline, 1998a, for a strong critique of any kind of entropy metric.)

In the spirit of the opening remarks in the Prologue—namely, that no demonstrably new science likely need be invented to understand cosmic evolution and its attendant rise in complexity—we prefer to return to a steadfast concept of fundamental thermodynamics and to characterize that complexity by using quantifiably straightforward terms. In short, energy and energy flow seem to be more accessible, explicit, and primary quantities, and not just because this is a worldview espoused from an admittedly physical perspective. What is more, the concept of energy remains meaningful for any macroscopic state, obviating the difficulties noted above for entropy or information. More than any other term, energy has a central role to play in each of the physical, biological, and cultural evolutionary parts of the inclusive scenario of cosmic evolution; in short, energy is a common, underlying factor like no other—a "DC baseline" in physicists' lingo—in our search for unity among all material things.

Recall from basic thermodynamics that the total energy, E, of a system equals the free energy, F, of that system plus the product of temperature, T, and entropy, S, characterizing the system. In symbolic form, from the Introduction, $E = F + TS$. Thus if, for a given temperature, the entropy of some system is to decrease—and this is the essence of order and organization—then the free energy must increase. Yet open-system, non-equilibrium thermodynamics is not concerned with the absolute value of a structure's total free energy as much as with its free energy density. After all, a galaxy clearly has more energy than a cell, but of course galaxies also have greater sizes and masses; it is the organized energy *density* that best characterizes the degree of order or complexity in any system, just as it was radiation energy density and matter energy density that were significant earlier in the Universe. In fact, what is more important is the *rate* at which free energy transits a complex system of given mass; accordingly, all systems can be compared on a fair and level spectrum. In other words, a most useful quantity used to specify operationally the order and organization of any system is the free energy rate density, alternatively called the specific free energy rate, expressed in units of energy per time per mass and denoted by the symbol $\Phi_m$. (This term differs from the "dissipative" function, $\Phi$, a quantity having dimensions of power in conventional non-equilibrium thermodynamics; compare Caplan and Essig 1985). Our term, $\Phi_m$, is familiar to astronomers as the luminosity-to-mass ratio, to physicists as the power density, to geologists as the specific radiant flux, to biologists as the specific metabolic rate, and to engineers as the power-to-mass ratio. We prefer to employ the more straightforward designation, *free energy rate density*, partly to emphasize its expressed physical meaning—a rate of energy flow, not a flux per se, and a mass density, not a volume density—and partly to stress its interdisciplinary application among all the natural sciences (Chaisson 1997a; 1998).

Actually, the idea of energy flow is not novel to an understanding of systems, especially ones displaying biological expression. Three-quarters of a century ago, biometrician Alfred Lotka (1922) identified energy's vital role, with "evolution . . . proceeding in such a direction as to make the total energy flux through the system a maximum compatible with the constraints" (see also Odum 1983). Today we would suggest that these systems *optimize* their flow. In turn, Lotka acknowledged his thermodynamic forebears: "It has been pointed out by Boltzmann that the fundamental object of contention in the life-struggle, in the evolution of the organic world, is available energy. In accord with this observation is the principle that, in the struggle for existence, the advantage must go to those organisms whose energy-capturing devices are most efficient in directing available energy into channels favorable to the preservation of the species." Lotka continues, even if not precisely according to today's improved standards, to stress the importance of energy flow: "In every instance considered, natural selection will so operate as to increase the total mass of the organic system, to increase the rate of circulation of matter through the system, and to increase the total energy flux through the system, so long as there is presented an unutilized residue of matter and available energy . . . and, so long as there remains an unutilized margin of available energy, sooner or later the battle, presumably, will be between two groups of species equally efficient, equally economical, but the one more apt than the other in tapping previously unutilized sources of available energy."

The biophysicist Harold Morowitz (1968), among others more recently (e.g., Jantsch 1980), has also emphasized the first-order consequences of energy flow, especially in biological systems, a subject to which we shall return in the Discussion: "If we were to surround the planet with an adiabatic envelope, all living processes would cease within a short time and the system would begin to decay to some equilibrium state. Solar energy influx is clearly necessary for the maintenance of life. It is just as necessary that a sink be provided for the outflow of thermal energy. If this were not so, the planet would continuously heat up and life would again soon cease to exist. All biological processes depend on the absorption of solar photons and the transfer of heat to celestial sinks."

Nor is the idea of energy itself a foreign concept in any transaction, least of all universal construction (Dyson 1971; R. Fox 1988; Smil 1999). Just about everyone would agree that this most basic of all physical quantities likely plays *some* role at virtually every level of development and evolution. But can the energy argument be made sound, integrated, and inclusive, applicable throughout all the sciences? Can we go beyond both the generalities and the biology, to explore some specific, quantitative aspects of energy, without losing sight of our proposed holistic synthesis? To simplify the analysis, indeed largely to err if need be on the side of the bigger picture and not the details, we consider energy pure and simple, ignoring for now subtle variations proposed by others, such as "exergy" (Eriksson et al. 1982), "emergy" (Odum 1988), "essergy" (Ayres 1994), or "endergy" (Corning and Kline 1998b), among sundry utility and efficiency factors affecting the type and quality of energy.

We also reject the occasional criticism that energy, however expressed and harking back to its classical thermodynamic underpinnings, is a nineteenth-century concept whose usefulness has come and gone. Admittedly an abstract idea invented more than a century ago to quantify the workings of Nature's many varied phenomena, energy nonetheless remains a vibrant, twenty-first-century term central to the understanding of probably all material things. One simply cannot examine a galaxy, a star, a planet, or any life form without taking energy seriously into account. Energy is the most universal currency in all of science, and we unapologetically employ it liberally in the subsequent analysis.

Let us begin by computing the energy flowing through a wide variety of structures representing a whole spectrum of perceived order. Here, we restrict our analysis in two ways. First, we examine only a handful of generic structures along the evolutionary path that has historically led to us, and also we consider only those structures as they exist now. Cosmic evolution is, after all, a proposition that seeks to understand who *we* are and whence *we* came. Later in the Discussion, our analysis is broadened, and strengthened, to include not only relevant structures at earlier times in *their* evolutionary track, but also representatives of those structures demonstrably ordered yet not within the cosmic evolutionary line of ascent that led to us. Knowing the size and scale of such structures, we can estimate the free energy rate density, $\Phi_m$, which should be a reasonable measure of the energy available to order a given amount of matter. As a reminder, the units are erg s$^{-1}$ g$^{-1}$, for again we stress that $\Phi_m$ is an energy rate (i.e., per unit time) density (i.e., per unit mass).

Consider first a star, in particular an average star such as the Sun. The solar luminosity today is $4 \times 10^{33}$ erg s$^{-1}$ and its mass is $2 \times 10^{33}$ g, making $\Phi_m$ for the Sun equal to 2 erg s$^{-1}$ g$^{-1}$. This is the average rate of energy release per unit mass of cosmic baryons which fuse approximately 10 percent of their hydrogen in one Hubble time (~10 billion years). This is also energy flowing effectively *through* the star, as gravitational potential energy during the act of star formation is now converted into radiation released by the mature star. Such a star, as analyzed at the end of the Introduction, utilizes high-grade energy in the form of gravitational and nuclear events to gain for itself greater organization, but only at the expense of its surrounding environment; it emits low-grade light, which, by comparison, is a highly disorganized entity. This is a relative statement, however: As we shall soon see, what we have called here "low-grade," disordering sunlight will, when later encountering Earth, become a high-grade ordering form of energy compared to the even lower-grade (infrared) energy re-emitted by Earth.[3]

A typical, normal galaxy has a value of $\Phi_m$ comparable to that of a normal star largely because, when examined in bulk, galaxies are hardly more than gargantuan collections of stars. Since the total energy of a galaxy scales roughly as the number of its stars, it has nearly a trillion times the luminosity of an average star. Yet its value of $\Phi_m$—an effective energy *density*—also scales inversely as the mass of the entire galaxy housing those stars, making $\Phi_m$ just a bit smaller than for stars. For example, our Milky Way Galaxy has a net energy flow of some $10^{45}$ erg s$^{-1}$ (including contributions from interstellar clouds and cosmic rays as well as stars), and a mass of roughly $2 \times 10^{45}$ g (including loose interstellar gas and dust in the galactic plane as well as anonymous dark matter thought present in the galactic halo beyond the plane). Such a resulting low value of $\Phi_m \simeq 0.5$ erg s$^{-1}$ g$^{-1}$ accords with our preconceived notion that galaxies, despite their majestic splendor, are not terribly complex compared to many other forms of organized matter.[4]

Planets are more complex than either stars or galaxies, and not surprisingly planetary values of $\Phi_m$ are larger—at least for some parts of some planets at some time in their history. (That's why we probably know more about the Sun than the Earth; stars are simpler systems.) We treat here not a planet's whole globe, from the interior through the surface, since the planets of our Solar System are not evolving much now, some 4.6 billion years after their formation; rather, we are more interested in those parts of planets that are still evolving robustly, still requiring energy to maintain (or regenerate) their structure and organization. Consider, for example, the amount of energy needed to drive Earth's climasphere, the most impressively ordered inanimate system at the surface of our planet today. The climasphere includes the lower atmosphere and upper ocean, which most affect meteorological phenomena capable of evaporating copious amounts of water as well as mechanically circulating air, water, wind, and waves. The total solar radiance intercepted by Earth is $1.8 \times 10^{24}$ erg s$^{-1}$, of which only about 70 percent penetrates the atmosphere (since Earth's albedo $\simeq 0.3$). This is several thousand times the power that Earth's surface now receives from its warm interior, which can thus be neglected, but we shall later return to examine Earth's radiogenic and accretional energies during its formative stage. Since our planet's air totals $5 \times 10^{21}$ g and that mixed layer, the upper 30 m of the oceans, engaged in weather comprises approximately double that, the value of $\Phi_m$ for planet Earth is roughly 75 erg s$^{-1}$ g$^{-1}$. The re-emitted infrared photons (equal in total energy to the captured sunlight) are both greater in number and lower in energy (about twenty times difference per photon) than the incoming sunlight of yellow-green photons, thus again continually contributing to the rise of entropy beyond Earth.[5]

Living systems require substantially larger values of $\Phi_m$ to maintain their order, including growth and reproduction. Photosynthesis is the most widespread biological process occurring on the face of the Earth, dating back at least 3.3 billion years when rocks of that age first trapped the chlorophyl porphyrins that drive this process in green plants, bacteria, and algae. These lower life forms need 17 kJ for each gram of photosynthesizing biomass, and they get it directly from the Sun. Since the annual conversion of $CO_2$ to biomass is $1.7 \times 10^{17}$ g (or ~10 billion tons of carbohydrates), the entire biosphere must use energy at the rate of nearly $10^{21}$ erg s$^{-1}$ (or about 0.1 percent of the total radiant energy reaching Earth's surface). And given that the total mass of the terrestrial biosphere (living component only) is approximately $10^{18}$ g, then $\Phi_m$ for the physico-chemical process of photosynthesis is roughly 900 erg s$^{-1}$ g$^{-1}$.[6]

Humans, by contrast, consume typically 2800 kcal per day (or 130 watts in the form of food) to drive our metabolism; this energy, gained indirectly from that stored in other (plant and animal) organisms and only indirectly from the Sun, is sufficient to maintain our body temperature and other physiological functions as well as to fuel movement during our daily tasks. Metabolism, by the way, is a genuinely dissipative mechanism, thus making a connection with previous thermodynamic arguments that some might have (wrongly) considered pertinent only to inanimate systems. Having an average male body mass of 70 kg, we therefore maintain a $\Phi_m$ of some $2 \times 10^4$ erg s$^{-1}$ g$^{-1}$ while in good health. This is how humankind contributes to the rise of entropy in the Universe: We consume organized energy in the form of structured foodstuffs, and we radiate away as body heat an equivalent energy in the form of highly disorganized infrared photons. We, too, are dissipative structures—highly evolved dissipators.[7]

In turn on up the complexity continuum, the adult human brain—the most exquisite clump of matter in the known Universe—has a cranial capacity of typically 1300 g and requires about 400 kcal per day (or 20 watts) to function properly. Our brains therefore have a $\Phi_m$ value of roughly $1.5 \times 10^5$ erg s$^{-1}$ g$^{-1}$. This large energy density flowing through our heads, mostly to maintain the electrical actions of countless neurons, testifies to the disproportionate amount of worth Nature has invested in brains; occupying 2 percent of our body's mass yet using nearly 20 percent of its energy intake, our cranium is striking evidence

*Table 2.* Some Estimated Free Energy Rate Densities

| Generic Structure | Approximate Age ($10^9$ y) | Average $\Phi_m$ (erg s$^{-1}$g$^{-1}$) |
| --- | --- | --- |
| galaxies (Milky Way) | 12 | 0.5 |
| stars (Sun) | 10 | 2 |
| planets (Earth) | 5 | 75 |
| plants (biosphere) | 3 | 900 |
| animals (human body) | $10^{-2}$ | 20,000 |
| brains (human cranium) | $10^{-3}$ | 150,000 |
| society (modern culture) | 0 | 500,000 |

of the superiority, in evolutionary terms, of brain over brawn. Thus, to keep thinking, our heads glow (in the far-infrared) with as much energy as a small lightbulb; when the "light" goes out, we die.[8]

Finally, we consider civilization en masse, the open system of all humanity comprising modern society going about its daily, energy-driven business. Today's ~6 billion inhabitants utilize ~18 trillion watts to keep our technological culture fueled and operating, admittedly unevenly distributed in localized pockets across the globe. The cultural ensemble equalling the whole of humankind then has a $\Phi_m$ value of some $5 \times 10^5$ erg s$^{-1}$ g$^{-1}$. Not surprisingly, a group of brainy organisms working collectively is even more complex than the totality of its individual components, at least as regards our criterion for order of free energy rate density—a good example of "the whole being greater than the sum of its parts," without resorting to anything other than the flow of free energy through an organized, and in this case social, open system.[9]

Table 2 summarizes the values of $\Phi_m$ for these seven representative structures spanning a wide spectrum of complexity, along with their specific cases computed above in parentheses. Also listed are the ages (in years) of each type of structure, dating back to their origins in the observational record; thus these are not ages of individual structures, rather the total duration of that type of structure's existence. As such, we do not mean to imply that the Sun is 10 billion years old; it clearly has only half that age, but stars generally have existed over the course of the past 10 billion years. Likewise, no given life form has an age of 3 billion years, but Earth's photosynthesizing biosphere has existed for at least that long. Figure 28 graphs the temporal gain in $\Phi_m$ among these open systems, illustrating its rapid rise in more recent epochs. The val-

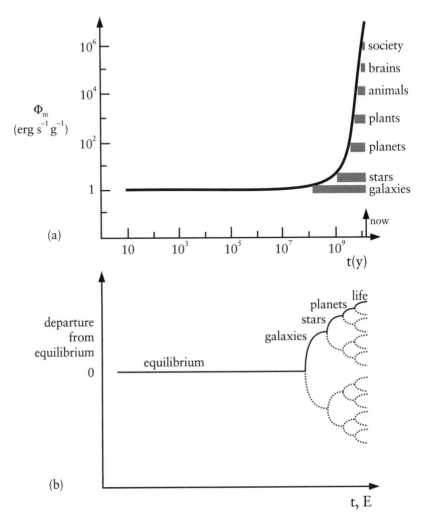

Figure 28. (a) The rise in free energy rate density, $\Phi_m$, plotted as histograms starting at those times at which various open structures emerged in Nature, has been rapid in the last few billion years, much as we suspected from human intuition and Figure 3 of the Prologue. The solid line approximates the upturn in negentropy or information content sketched in Figure 27(b), but as graphed here it is energy flow that best characterizes order, form, and structure in the Universe. (b) This qualitative analog of the quantitative plot in (a) makes a connection with the earlier discussion (Figure 9) of systems' chancy fluctuations responding deterministically to environmental conditions, which in turn gave rise to a hierarchy of complex states via bifurcating changes. The solid lines represent the paths taken historically, the dotted lines options never realized. This does not imply that one type of structure evolves into another; rather, new and more complex structures likely emerged as energy flows became more prevalent yet more localized.

ues of $\Phi_m$ are plotted as horizontal histograms, starting at the times when each of these different systems first appeared in Nature and extending to the present, for not only has each type of generic structure originated at the times given in the table but they have also endured since then.

Clearly, $\Phi_m$ has increased steadily as more intricately ordered structures have emerged, in turn, throughout cosmic history, indeed dramatically so in relatively recent times. Although the total energy flowing through a star or planet is hugely larger than that through our human body or brain, the *specific* rate—in effect, the rate *density*—is much larger for the latter. Thus, organized systems observed in the Universe do in fact temporarily increase their information content, or complexity, in actuality $\Phi_m$ faster than I, no doubt because factors such as functional traits in biological evolution grant distinct advantages to those systems able to process additional free energy over and above that provided by thermal gradients engendered by the expanding Universe.

That the temporal dependence of the free energy rate density (the solid curve in Figure 28) mimics the suspected rise of complexity sketched earlier (Figure 3) is not to be overlooked. If this extended essay reasonably approximates reality, then the two are the same, or nearly so. Significantly, one explains the other (again, or nearly so), thereby reaching the goal we set out to achieve at the start of this book. This is not to say, by any means, that galaxies per se evolved into stars, or stars into planets, or planets into life. Rather, our analysis suggests, and not just qualitatively in words but with coarse quantitative reasoning, that galaxies gave rise to environments suited for the birth of stars, that some stars spawned environments conducive to the formation of planets, and that an untold number of planets fostered environments ripe for the emergence of life.

Take another close look at three curves: Figure 3 graphs our "gut feeling" of growing complexity as a function of time, not knowing when we began whether that quantity might have risen linearly, stepwise, exponentially, or in some other way. Figure 27(b) shows how the cosmic conditions in an expanding Universe generate potential for the growth of complexity, although that curve derives largely from theory laden with assumptions. And Figure 28(a) displays the data for $\Phi_m$, however limited a measure of true complexity this term might be (consult the next few paragraphs). This last curve clearly rises more sharply than either of the first two curves suggest, though this should not surprise us. As will be noted in the upcoming Discussion, Darwinian biological evolution and Lamarckian cultural evolution, among other catalytic effects during physical evolution, might well have fostered additional complexity on local scales beyond that possible by universal expansion alone. That's probably why, in a relative sense, physical evolution is sluggish, biological evolution moderate, and cultural evolution rapid; $\Phi_m$ is a kind of motor of evolution (or at least its fuel), accelerating systems' abilities to assimilate increased power densities.

Nor should we be puzzled by the shape of the $\Phi_m$ curve in Figure 28; there is nothing peculiar about the extraordinary rise in the most recent billion years or so. The mathematical nature of any exponential curve—much like that for human population on Earth—causes it to increase initially only very slightly and slowly over long time intervals (going, for example, from 1 to 2 percent in one doubling period), after which it surges upward nearly explosively during a short duration (going from 50 to 100 percent also in a single doubling period). There need not be any designer agents or mystical acts, yet unidentified, responsible for the dramatic increase in $\Phi_m$ values; the natural phenomena described herein seem quite capable of accounting for it.

In calculating values of free energy rate density and in drawing time-dependent graphs of it, two caveats need to be underscored—one philosophical and the other technical. First and foremost, in no way whatsoever is the rising curve of $\Phi_m$ in Figure 28 meant to claim or imply anthropocentrism. That humankind and its social inventions comprise the greatest complexities known in Nature seems indisputable; that we harbor the most intricate bundle of matter observed anywhere is presently impossible to deny. Yet no inference is construed about humanity epitomizing any pinnacle or culmination of the evolutionary process. No hidden, anthropocentric agenda lurks here. We are neither the centerpiece nor the final product of this remarkable cosmic-evolutionary narrative, even though we *are* the only known structures conscious of that story. To be sure, it does not follow that some straight and narrow evolutionary path led directly from big bang to humankind; rather, that path, extending over some 12 billion years, was contorted at best, with many failures and extinctions among innumerable sidetracks. The empirically derived plot of $\Phi_m$, resembling our gut feelings about complexity expressed in the Prologue, merely captures the highlights of the salient events that did eventually produce a hierarchical distribution of many evolved systems, in effect representatives of the "winners," without showing the countless failures. In fact, disorder is generally more probable than order largely because so many more paths to disorder prevail, much as there are always more ways to make nonsense than sense. Figure 28 and Table 2 together comprise one way of putting all known structured systems on a common technical ledger, which was one of our objectives from the outset. The curve of complexity is still rising, but to what end—if there is an end—contingent science cannot say. Nor, despite much development and evolutionary advancement in natural history, do we see any evidence for obvious design or overt purpose.

Deep-seated anti-anthropocentrism is the root cause of most physicists' and astronomers' inclination toward a committed search for extraterrestrial intelligent life. Just as our telescopes probe cosmic matter and radiation in all (or many) of its manifestations, we seek to take inventory of all (or many) life forms in the Universe, partly for completeness and partly to test the proposition that our intelligence is magnificently mediocre. The analysis presented here—that the temporal rise of free energy rate density as a measure of growing complexity ultimately derives from universal expansion—augurs well for the emergence of life at myriad localized sites in the Universe. With continued cosmic expansion, that analysis furthermore implies the onset of ever more complex entities, though not necessarily here on Earth. By contrast and ironically, it is the biologists and anthropologists who generally oppose such a search for extraterrestrials, or at least think unwell of it, claiming that the chances for intelligent life elsewhere are negligible—an attitude reminiscent of the now-discredited metaphor of *Homo sapiens* atop the evolutionary tree, much like an angel ornament perched on a Christmas tree. The urge to search ought to prevail, if only by way of negative argument: By failing to search in some modest way for extraterrestrial life, we commit the cardinal sin of pre-Renaissance workers—just talking and debating, like philosophers of antiquity, without experimentally or observationally testing the age-old hypothesis of life beyond Earth.

A second caveat concerns the level of detail in our computational analysis; to be honest, we have skirted some of the hardest details. In particular, as noted at the outset of our calculations, the values for $\Phi_m$ employ only bulk energy flow, that is, total energy available to a handful of representative open systems. Accordingly, quantity, or intensity, of energy has been favored while largely neglecting measures of quality, or effectiveness, of that energy. Clearly, a more thorough analysis would incorporate such factors as temperature, type, and variability of an emitting energy source, as well as the efficiency of a receiving system to use that free energy flowing through it. After all, input energy of certain spectral wavelengths can be more useful or damaging than others, depending upon a system's status, its receptors, and its relation to the environment. Likewise, the efficiency of energy use can vary among systems and even within different parts of a given system; under biological conditions, for example, only some of the incoming energy is available for work, and technically only this fraction is the true free energy. That energy might benefit some parts of a system more than others is a necessary refinement of the larger opus to come. For this abridgment, our estimates suffice to display general trends; the next step is a more complete (perhaps we should say more "complex") study to examine more closely how, and how well, open systems utilize their free energy flows to enhance complexity.

Even the absolute quantity of energy flowing through open systems needs to be more carefully considered in a detailed analysis. Not just any energy flow will do, as it might be too low or too high to help complexify a system. Very low energy flows mean the system will likely remain at or near equilibrium with the thermal sink, whereas very high flows will cause the system to approach equilibrium with what must effectively be a hot source—that is, damage the system to the point of destruction. That much was clear for the classically dissipative Bénard cells encountered late in the Introduction. Sustained order is a property of systems enjoying moderate, or "optimum," flow rates; it's a little like the difference between watering a plant and drowning it.[10] In other words, a flame, a welding torch, and a bomb, among many other natural and human-made gadgets, have such large values of $\Phi_m$ as to be unhelpful. Take, for example, a simple flame:[11] An ordinary candle is clearly a dissipative, kinetic structure when lit, its dynamic steady state maintained by a continuous flux of matter (incoming $O_2$, outgoing $CO_2$) as the mass of the candle dissolves by burning at the wick. Given its concentrated luminosity yet low mass, a flame unsurprisingly has a very high $\Phi_m$ value (5 W output for a few tens of grams, or $\sim 10^6$ ergs $s^{-1} g^{-1}$), which is probably why some candles, when also unchecked like bombs, can often be so harmful; their energy flows are too large, thereby driving their contents to destruction, indeed toward unstructured equilibrium. "Order, we must have order—but not too much order!" said A. N. Whitehead nearly a century ago, another champion of process and reality rooted in energy flow. We return to this issue early in the Discussion when encountering supernovae, which are clearly objects of extremely high energy rate density, precisely as expected for one of the most unconstructive events in Nature.

As a summary illustration of this book's three chapters, Figure 29 graphs a superposition of the temporal dependence of $aT^4$, $\rho_m c^2$, and $\Phi_m$, thereby delineating the variation of radiation, matter, and complexity throughout all history. (Note that this comparison, at this point in an ongoing research program, is only aesthetic since $\Phi_m$ has units different from those of the other two energy densities; $\Phi_m$ is also a localized quantity pertaining to structures specifically, whereas the other energy densities are global quantities pertaining to the Universe generally.) Whereas the intersection of the energy densities of matter and radiation denotes the transformation of the Radiation Era into the Matter Era, the ascent of free energy rate density over both these energy densities heralds a genuinely new era—the Life Era. In particular, the controlled use of radiation by animated structures (for instance, when the free energy density exceeds the radiation energy density, $\Phi_m > aT^4$, beginning with a simple cell) is identified with one of the first great inventions of biological evolution: photosynthesis. This is a grand event whereby life comes to dominate radiation, utilizing sunlight in a survival-related fashion. An even grander event occurs when life forms dominate matter, that is, when the free energy density exceeds the matter energy density, $\Phi_m > \rho_m c^2$, beginning with technological intelligence on Earth and possibly elsewhere. *Only the latter of these two changes denotes the onset of the Life Era,* because only with the origin of technologically manipulative life, not just life itself, does life exert leverage over both radiation *and* matter. (Gaian microbes do not signify the start of the Life Era since, although they may well regulate the environment, they do not literally manipulate it, nor in any way control the genes as only technologically advanced life can.)

The transition from the Matter Era to the Life Era will not be instantaneous; it is an evolutionary, not revolutionary, phase change. Just as much time was needed for matter to conquer radiation in the early Universe, long durations will likely be required for life to best matter. Life might not, in fact, ever fully dominate matter, either because civilizations fail to gain control of material resources on truly galactic scales or because the longevity of technological civilizations everywhere is inherently small.

## Summary of Chapter Three

The sources and sinks of energy flows through complex, yet conatural, entities such as stars, planets, and life all relate back to the time of ther-

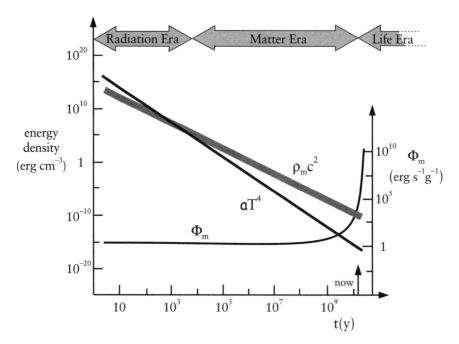

Figure 29. A plot of the temporal behavior of the principal energy densities delineates and distinguishes the three major eras in the history of the Universe.

mal symmetry-breaking in the early Universe. This is history's preeminent change from the Radiation Era to the Matter Era, a period when the cosmic conditions naturally evolved to foster the emergence of order and organization, and increasingly so with the expansion of the Universe ever since. More than any other single factor, energy flow is the principal means whereby all of Nature's diverse systems naturally generate complexity, some of them evolving to impressive degrees of order characteristic of life and society. Furthermore, for all structured systems, entropy increases of the larger surrounding environment can be mathematically shown to exceed the entropy decreases of the localized systems per se, guaranteeing good agreement with the second law of thermodynamics. We thus arrive at a clean, clear reconciliation of the theoretical destructiveness of thermodynamics with the observed constructiveness of cosmic evolution.

Although a mature Life Era may never fully come to pass, one thing seems certain: Our generation on planet Earth, as well as any other neophyte technological life forms populating the Universe, are currently participating in an astronomically significant transformation, a second magisterial change in the history of the cosmos. We now perceive the dawn of a whole new reign of cosmic development—an era of opportunity for advanced life forms to begin truly to unlock secrets of the Universe, to fathom our role in the cosmos, indeed to decipher who we are and whence we came. We have become smart enough to reflect back upon the material contents that gave us life, and the result is a rich natural history unparalleled in human knowledge.

# EVOLUTION, BROADLY CONSIDERED

That living organisms have evolved on Earth, there is no doubt. Evolution is as much a fact as is the Earth revolving around the Sun. Well more than a century ago, Charles Darwin proposed a genealogical connection among all organisms and a history of life regulated by "descent with modification"—a working idea that all organisms and lineages have descended from a simple ancestral form, to be sure an idea now accepted by virtually all active scientists. What we scientifically debate today are the ways, means, and rates of evolution, the mechanisms of change. Is there an underlying action driving evolution and is natural selection the only process guiding it? Are specific environmental conditions needed for change and are those conditions in accord with known science? What about a "life force"; must we seek novel insight, something akin to vitalism or *vis vitae*, for the emergence and maintenance of life on Earth? The answers suggest that no demonstrably new science is needed, provided we are willing to think innovatively and expansively about evolution itself.

The word *evolution* need not be the sole province of biology, its usefulness of value only to life scientists. From the very first edition of his greatest work, *On the Origin of Species* (1859), Darwin never actually used that word as a noun, and only once, in the very last word of the book, did he use it as a verb: "There is grandeur in this view of life, with its several powers, having been originally breathed into a few forms or into one; and that, whilst this planet has gone cycling on according to the fixed law of gravity, from so simple a beginning endless forms most beautiful and most wonderful have been, and are being, evolved."

Nor need we restrict ourselves to the principle of natural selection as the sole cause of evolutionary change, past and present. Natural selection has surely been a prominent mechanism (indeed likely *the* major factor) in the evolution of life on Earth, especially of the resulting adaptation of an organism's form, function, and behavior to achieve enhanced reproductive success. Only recently, however, has natural selection come to be seen by some as ruling biological evolution absolutely. The power of this quintessential Darwinian concept is so pervasive in biology and related fields that many so-called ultra-Darwinists subscribe to it wholly and without limit, indeed with a kind of fervor matched only by the fundamentalists whom they abhor. Straight and narrow biologists have adopted natural selection as an uncompromising ideology based on genes ruling all life (Dawkins 1996), most evolutionary psychologists (née sociobiologists) are using it to revolutionize the study of human behavior along strict Darwinian lines (Wright 1994), and even a few philosophers now appropriate it to explain the meaning of life and consciousness (Dennett 1995). Yet even in Darwin's time, he eventually felt obliged, in the preface to the last edition (1872) of his classic, to make clear that life's many varied changes were not likely rendered by any single cause, however ubiquitous and powerful: "As my conclusions have lately been much misrepresented, and it has been stated that I attribute the modification of species exclusively to natural selection, I may be permitted to remark that in the first edition of this work, and subsequently, I placed in a most conspicuous position—namely at the close of the Introduction—the following words: 'I am convinced that Natural Selection has been the main but not exclusive means of modification.' This has been of no avail. Great is the power of steady misrepresentation."

In contrast to the ultra-Darwinists, the Darwinian pluralists argue that natural selection alone cannot suffice to explain many aspects of evolution; other mechanisms are relevant, perhaps even prevalent, outside Darwin's traditional premise that selective effects act mostly and gradually on groups of individuals within populations of organisms (Gould 1982). The pluralists contend that *some* structure emerges among life forms without recourse to gradualistic natural selection, or by some other means in consort with it. Not that they deny either the existence or importance of natural selection, for its validity has been demonstrated by countless observations and experiments. Rather, the pluralists aim to identify powerful adjuncts to natural selection, to include additional mechanisms and causal elements contributing to the history of life, such as rapid (punctuated-equilibrium) changes that cause branching off of whole new populations of species and not just individuals (Eldredge and Gould 1972), evolution via neutral (non-adaptive, drifting) changes among the nucleotide bases of DNA comprising the core of population genetics (Kimura 1983), and the spontaneous (self-organized) changes of molecules at critical energy thresholds within many complex systems (Kauffman 1993; Bak 1996). These and other enriching evolutionary schemes, seemingly "in the air" at the end of the twentieth century, concern mostly life and its manifold changes; yet they remain parochial given the broader landscape sketched here.

Actually, the term "selection" is itself a bit of a misnomer, for there is no known agent in Nature that deliberately selects. Selection itself is not an active "force" or promoter of evolution as much as a passive pruning device to weed out the unfit. As such, selected objects are simply those that remain after all the poorly adapted or less fortunate ones have been removed from a population of such objects. A better term might be "nonrandom elimination" (Mayr 1982; consult note 1 in the Introduction), since what we really mean to characterize is the aggregate of adverse circumstances responsible for the deletion of some members of a group. Accordingly, natural selection can be broadly taken to mean preferential interaction of any object with its environment. This more liberal interpretation also helps widen our view of evolution, an admittedly key objective of the present work.

Selection is a factor in the flow of energy into and out of all open systems, not just life forms; in short, systems are selected by their ability to command energy resources—and this energy is the "force," if there is any at all, in evolution. Broadly considered, selection does occur in the inanimate world, often providing a formative step in the production of order. For example, as warm air attempts to escape from a room, it might drain through a crack under the door, leak through a poorly insulated window, or slowly conduct through the walls. The physical geometry and environmental conditions select the path for the energy flow; if the crack under the door is large, it will rapidly dominate the other two paths, and the outflowing energy will soon return the room to equilibrium with its surroundings. The net change is not often large, as such a room is usually never far from equilibrium. However, as noted in the Introduction, when the disequilibrium state is substantial, the energy flow can exceed some critical threshold and selection, in turn and especially at bifurcation points, can become part of the process of generating newly ordered forms. This, again, is what spontaneous organization is all about.

Toward the end of the Introduction, we encountered several other examples of inanimate selection, without stressing it as such, and always in the presence of energy. Certain modes of radiation able to utilize energy coherently are "selected" for laser propagation, rare atmospheric eddies able to enhance energy flow are "selected" to become storms, among other examples soon to be considered. The interstellar medium provides yet another example of selection naturally operating among inanimate objects, as pre-biological molecules undergo chemical ordering, again the transaction always involving energy. Here, under very low density conditions ($\rho_m \simeq 10^{-24}$ g cm$^{-3}$), those molecules that are relatively stable to ultraviolet (UV) radiation are "selected" to attain eventually an abundance higher than their relatively small production rates might suggest over the course of generations of century-long cycles of formation, destruction, and reformation; these are among the organic molecules found by radio astronomers in vast galactic clouds. What's more, electrons in certain heterocyclic (mostly carbon-ringed) molecules are able to absorb UV as an excitation rather than as a dissociation event much more readily than can simpler molecules, implying that, even in the harsh realities of interstellar space, the high molecular-weight cyclic compounds might be favored, once formed; these are among the organic molecules found in carbonaceous-chondrite meteorites (Sagan 1973).

In this brief monograph, designed to give merely the gist of an advancing research program, we have discussed the process of evolution as it applies to much more than life on Earth. We have taken the liberty of using the word "evolution" in an intentionally provocative context, to capture ontological, ecological, and phylogenetic change on all spatial and temporal scales by means surely including, but not restricted to, natural selection. We have sought, within the expansive scenario of cosmic evolution, general trends among Nature's myriad changes over the course of an impressively long line of temporality, from big bang to humankind. And we have been especially alert to any changes—developmental or generative, gradual or punctual—in the universal environment that might have allowed for, indeed driven, evolution from time immemorial.

Critics will contend that cosmic evolution is just another faddish "c-theory," given history's recent tendency to search for a comprehensive, underlying, neo-Platonic order that has regularly eluded science. Like creation-field cosmology of the 1940s, cybernetics ideas of the 1950s, catastrophe models of the 1960s, chaos theory of the 1970s, and complexity science of the 1980s, perhaps cosmic evolution is merely the latest, 1990s version of a futile attempt to develop a unified view of Nature. Yet cosmic evolution does seem to have in its favor several promising features for a scientific synthesis at the dawn of the new millennium:

- a vast body of experimental and observational data, accumulated throughout all the natural sciences and organized into a unifying, coherent framework of understanding,

- an impressive array of animate and inanimate objects whose ages, energies and apparent complexities sequence well along a cosmological arrow of time,

- a cogent worldview that derives from known science, especially non-equilibrium thermodynamics, without any need to postulate new kinds of science or mysticism,

- a synthesizing scenario that readily incorporates a mixture of chance and necessity, robustly explaining much of the past without claiming to predict any of the future,

- a sweeping, epic-class story of rich natural history having mythological proportions that everyone can appreciate and comprehend.

### Evolution: A Physical Perspective

Biologists and geologists, among other Earth-oriented scholars, are understandably reluctant to look beyond our home planet for answers to some of the secrets of change—to widen and deepen their vista to include a more eclectic view of evolution in the astronomical realm, indeed to admit the kind of grand-scale change in which life could be a mere, albeit important, part. They have always been so reticent, even when it became clear during the past decade, for example, that asteroidal impacts have almost certainly affected life on Earth, a relatively recent one having apparently landed some 65 million years ago in the Yucatan and wiped out nearly half of the world's species and not just the dinosaurs. What is more, cosmic rays from well beyond Earth are now understood to be at least one of the chief causes of mutations—a veritable motor of evolution—that alter the makeup of life's genetic code. Still, most terrestrial scientists look inward (for that is rightfully their subject matter), when we all ought to be looking outward as well as inward, thus creating a more balanced view that allows for evolution by means perhaps only partially understood and on scales a good deal larger than heretofore widely recognized.

If any fundamental phenomenon underlies the development of order, form, and complexity all around us, it would seem to be nothing more, yet nothing less, than the expansion of the Universe. As suggested in the last chapter, the organized dynamics of the Universe itself are a necessary, if not sufficient, condition for the emergence of structured systems. Those dynamics established the temperature gradients, began the energy flowing, and fostered environmental changes literally everywhere. Yet cosmic expansion is not likely the only source of order and complexity in the Universe. Superposed on the primal, gradual changes in the early Universe that set the stage for the rich hierarchy of complex events colloquially termed earlier the "cosmic change of being" are other, more sophisticated mechanisms of change. Gravitational force in physics, natural selection in biology, and technological innovation in culture are all examples of diversified actions that can give rise to accelerated rates of change at locales much, much smaller than the Universe per se—such as the islands of order called stars, planets, and life itself.

On moderate scales, the evolution of gravitationally bound systems, for example, can diminish entropy and generate complexity, all the while entropy grows in the immense surrounding environment. An ordinary star is a good case in point. Such astronomical objects originate from tenuous pockets of gas and dust within otherwise chemically and thermally homogeneous galactic clouds—regions virtually in (or very near) equilibrium. A protostar poised at the verge of stardom is only slightly out of equilibrium, having a relatively small temperature gradient from core to surface and normally composed of a roughly uniform mixture of 90 percent hydrogen and 9 percent helium, often peppered with trace amounts of heavier elements. As the star ignites at the core, emerges from its placental envelope of mostly dust and begins to evolve physically, gravity drives it further from equilibrium;[1] the star's core steadily increases in temperature while adjusting its size like a cosmic thermostat, all the while nuclear fusion reactions change its lightweight elements into heavier types. With time, then, such an object grows thermally and chemically inhomogeneous, gradually becoming more ordered and less equilibrated; its core-surface thermal gradients increase as do its onion-like layers of fused heavy elements. While enhancing the structure of a star, such stellar evolutionary events inevitably create and store complexity, for a complete description of a thermally and chemically differentiated system requires more information than an equally complete description of its initially homogeneous state.

Can these statements be bolstered with quantitative analyses, relying not solely on qualitative reasoning, however sound? Indeed they can, at every stage of a star's evolutionary cycle. To give just one example, consider our Sun's principal fusion event, now underway in its core. This is the conversion of hydrogen ($H^1$) into helium (He), helped along by deuterium ($H^2$). The net nuclear (not chemical) reaction is

$$H^1 + H^2 \rightarrow He^3 + \gamma.$$

The photon ($\gamma$) results from the transformation of mass into energy of value $E_\gamma = 4.4$ Mev, or $7 \times 10^{-6}$ erg. This is a gamma-ray photon that can be regarded as an energy flow in the system, ultimately accruing from the gravitational potential energy of the parent galactic cloud and now carrying with it a change in entropy $\delta S = \delta Q/T = k_B(E_\gamma/k_B T)$. Here, the temperature in the Sun's core, $T \simeq 1.5 \times 10^7$ K, and the ratio, $E_\gamma/k_B T$, is effectively the number of thermalized photons, $N_{th}$ (i.e., average number of photons at the ambient temperature), equivalent to that one high-energy photon. In astronomical systems, numbers are so large that entropy approximately equals $k_B N_{th}$, because for an ideal gas, which the Sun closely resembles, the log factor in the entropy equation is slowly varying and of order unity; it is the factor N in the equation, $S = Nk_B \ln W$, that completely dominates. Thus, for the nuclear reaction above, the entropy of the two reactants on the left side is of order $2k_B$ per reaction, whereas the entropy of the products on the right is approximately $(1 + E_\gamma/k_B T)k_B = 3400$ $k_B$ per reaction. Therefore, on the one hand the spatial order of the Sun's system increases since He is deposited and concentrated at the core thereby creating an ordered structure that does not diffuse away, while on the other hand the value of the net entropy also increases in the Sun's surrounding environment since those photons eventually do indeed dissipate outward, as degraded sunlight. (Perspective is crucial, however: 8 minutes later, life on

Earth makes use of those photons, which though low-grade relative to the Sun's core are very much high-grade relative to photosynthetic processes on our planet. At issue is where in the to`. l energy budget events are examined, as here the Sun's "pollution" becomes life's vital resource.) Thermodynamics is intact: The solar environment is regularly disordered, all the while order emerges, naturally and of its own accord, within the stellar system.

Changes within and among stars can be cast into the same thermodynamics context developed at the end of Chapter Three. Putting aside measurements of negentropy or calculations of information content as both tricky and unproductive, we again appeal to the physically meaningful quantity, $\Phi_m$, free energy rate density. Values of $\Phi_m$ then serve as indicators of complexity, allowing the evolution of a star to be tracked as its interior experiences countless cycles of nuclear reactions that repeatedly cause the star to change size, color, brightness, and chemical composition. In this way, a star's passage from "birth" to "death" can be followed by noting its growing complexity through middle and old age, culminating in a return to the simplicity of equilibrium at the end of its productive "life" as a star.

Of course, stars for the most part change very much more slowly than the life spans of humans studying them. Astronomers are often portrayed as anthropologists, examining stars like bones of different ages, sizes, and makeups, and reorienting them like puzzle pieces while constructing a coherent natural history of stars. This is the domain of stellar evolution, one of the better-understood phases of the cosmic-evolutionary scenario. Figure 21 already summarized the most powerful tool in the graphical lexicon of a stellar evolutionist—the so-called Hertzsprung-Russell diagram—and it might be useful to consult it regarding the points made in the next few paragraphs.

Consider a few representative examples of stars at different stages in their evolutionary cycle. We have already calculated a $\Phi_m$ value of 2 erg $s^{-1}$ $g^{-1}$ for our Sun, which is a normal, average star in the midst of the main sequence of all stars. In some 5 billion years, the Sun will swell to become an ordinary red giant, much like the prominent star Arcturus, which has already achieved such a moderately complex state. With a luminosity L $\simeq$ 115 $L_\odot$ and a mass M $\simeq$ 5 $M_\odot$, Arcturus has $\Phi_m \simeq 45$ erg $s^{-1}$ $g^{-1}$. For a short period of time (yet still on the order of millions of years), our Sun might ultimately make its way partly up into the supergiant domain (see Figure 21), where today the bloated red-giant

Betelgeuse exemplifies an aged star already there; Betelgeuse has an even higher $\Phi_m$ value of about 350 erg s$^{-1}$ g$^{-1}$, based on its known energy output of $1.4 \times 10^4$ L$_\odot$ and an estimated mass of 80 M$_\odot$. Notice how these values of $\Phi_m$ steadily rise as the star passes through more advanced evolutionary stages in its lifecycle.

At the core of every red-giant star is one of the most ordered structures of inanimate matter to have evolved in Nature—a white-dwarf star. Several such dwarfs have been spotted in our galactic neighborhood, such as Sirius B (the compact companion to the brightest star in the nighttime sky, Sirius A), its red-giant envelope having receded into the surrounding interstellar medium. The theory of stellar evolution implies that the core of our low-mass Sun is fated to become such a white dwarf, a very dense ($\sim 10^7$ g cm$^{-3}$) ball of mostly carbon embers having dimensions about the size of Earth. As its mass is too small to generate further nuclear fusion, the center of a white dwarf compresses its constituent ions, ripping electrons from their atomic orbitals. The resulting sea of electrons, held buoyant not by heat but only by the Pauli exclusion principle and known technically as a degenerate Fermi gas, assumes a structural form displaying a kind of crystalline pattern, in fact increasingly so with age. Despite their high temperatures ($\sim 10^8$ K) and owing to their immense pressures, the strange cores of such dead stars are thought to resemble, ironically and largely due to gravity, the geometry of highly ordered crystals attainable on Earth only at the lowest temperatures (Lavenda 1985). Comparison of values of $\Phi_m$ for white dwarfs with those of genuine stars are inappropriate since these dwarfs are already "dead"; such balls of carbon ash are no longer fusing, no longer still ordering, no longer stars per se. White dwarfs are indeed like crystals; both are highly ordered yet without current energy input, although there is no doubt that the processes that created them most definitely displayed significant energy flows.

Low-mass, red-dwarf stars still on the main sequence are in fact fusing hydrogen into helium, albeit meekly, and here we can legitimately make comparisons to main-sequence stars. These kinds of dwarfs, including for example the well-known Barnard's star and Proxima Centauri of M $\simeq 0.1$ M$_\odot$, never attain red-giant status; rather, they proceed directly to the white-dwarf stage, but only after trillions of years of main-sequence burning. Unlike the Sun, wherein convection occurs only in the outer part of its layered gaseous ball, the low-mass red dwarfs are fully convective throughout their innards, resulting in much mixing, uniform composition, and no real layering. Little wonder, then, that such homogeneous stars have inordinately small values of $\Phi_m =$ $10^{-2}$ L$_\odot$/0.1M$_\odot \simeq 0.1$ erg s$^{-1}$ g$^{-1}$, much as expected for low-ordered, inanimate objects.

By contrast, some stars, though only those much more massive than the Sun (in fact at least several solar masses), eventually create great thermal and chemical gradients between their cores and surfaces, thereby reaching new levels of complexity. They do not last long, however, for when their internal temperature reaches $\sim 3 \times 10^9$ K and their composition becomes mostly iron within their inner cores, they catastrophically implode and then rapidly (probably within seconds) explode as supernovae. The most recent such event nearby, and the best studied supernova in history, was the explosion of the 30-M$_\odot$ star, Sk-69°202, whose intense flash of light first reached Earth in 1987. Now known as SN1987A, its internal complexity was presumably maximized (for a normal star) just prior to detonation, when, we calculate, this 10,000-solar-luminosity object would have had $\Phi_m \simeq$ 600 erg s$^{-1}$ g$^{-1}$.

This is not to say that indefinitely high values of $\Phi_m$ would in every case lead to ever more complexity. Examples abound in Nature where too much energy flow triggers just the opposite—namely, an open system that suffers breakdown, robbing the system of complexity and often returning it to equilibrium. At the end of Chapter Three, we noted how flames, bombs, and similarly destructive energy sources can effectively reverse the evolutionary process. A supernova event is another such case, wherein at the moment of explosion $\Phi_m$ is so large ($>>10^6$ erg s$^{-1}$ g$^{-1}$) as to blow the previously ordered star to smithereens. Such a violent event does in fact resemble a bomb; although the scales are vastly different, the values of $\Phi_m$ are similar and both results catastrophic. A 1-Megaton nuclear device, for example, has a huge $\Phi_m$ value of $\sim 10^{11}$ erg s$^{-1}$ g$^{-1}$, which explains its utter destructiveness. Such a bomb is uncontrolled, but someday when humankind masters nuclear fusion we might be able to control for positive evolutionary purposes such huge values of free energy rate density, as is the case already on miniaturized scales for modern computer chips, as noted at the end of this Discussion.

Figure 30 plots values of $\Phi_m$ for stars of varying complexity, including the case of a young Sun several billion years ago, when its luminosity and hence $\Phi_m$ were about a third of its present values.[2] Free energy rate density remains a useful way to quantify the growth of complexity, here during stellar evolution as elsewhere.

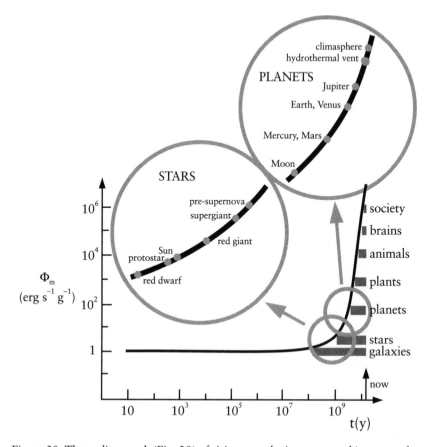

Figure 30. The earlier graph (Fig. 28) of rising complexity, expressed in terms of $\Phi_m$, the free energy rate density (reproduced at bottom), is here examined in more detail to highlight some of the ordered stages for a variety of inanimate objects discussed in the text—in this case, stars and planets in the physical-evolutionary phase of cosmic evolution.

Even more organized states of inanimate matter are reasoned to exist in the compressed kernels of neutron stars, those rotating remnants (pulsars) of massive stars that occasionally survive supernova explosions. These small, perhaps still hot objects of extreme density—mere tens of kilometers across yet with huge densities on the order of $10^{13}$ g cm$^{-3}$—comprise virtually pure neutrons, the remnants' electrons having collided with protons to yield a highly ordered neutron sea that effectively competes with gravity. Here, for stellar cores of more than 1.4

$M_\odot$ (the "Chandrasekhar limit"), the onslaught of gravity is enough to overwhelm the Pauli principle for electrons but not that for neutrons, thereby causing the core to collapse to even smaller size and greater density than for a white dwarf. Again ironically, such an uncommonly ordered neutron star, having a density comparable to an atomic nucleus, might well resemble a superconducting fluid like that displayed by liquid helium at some of the lowest temperatures ($10^{-3}$ K) attainable in the laboratory (Careri 1984).

Given that stars are demonstrable energy (and entropy) converters, they thus represent relatively localized sites of growing complexity—first, because stars also *radiate* entropy (as well as energy) into their surrounding environment and, second, because the gravitational agents tending to enhance stellar gradients usually overwhelm the opposing, non-gravitational agents (like heat) tending to diminish them. In essence, stars are islands of increasing negentropy in a universal sea of rising chaos, and much like most other ordered structures, they obey, within wide range and variation, our reasoned criterion of free energy rate density as a measure of order, flow, and complexity in the material world.

Ponder one more point regarding open, self-gravitating systems—which, by the way, stars are. Such a "gas without walls" is quite unlike any isolated system in a laboratory. Stars, among other members of the cosmic hierarchy of material clusters, do not readily relax toward a state of thermodynamic equilibrium with their surroundings. Even the terminal phase of many stars may not be, at first, an equilibrium state. Instead, their fate is probably one of free gravitational infall. For when a very massive star (>2.5 $M_\odot$, the "Landau-Oppenheimer limit," a lower limit in the absence of rotation) loses its previously maintained hydrostatic equilibrium between inward-pulling gravity and outward-pushing gas, it collapses catastrophically toward the bizarre configuration known as a black hole. The British astrophysicist, Martin Rees (1988), described it thus: "Gravitational binding energy is being released as stars, galaxies, and clusters progressively contract, this inexorable trend being delayed by rotation, nuclear energy, and the sheer scale of astronomical systems, which make things happen slowly and stave off gravity's final victory." If black holes ultimately gobble up all matter, with the increasing areas of their widening event horizons proportional to thermodynamic entropy (Thorne 1994), then they too can be said to tend toward an obligate equilibrium as Nature's premier states of maximal entropy for any given space. Even if black holes evaporate—though no observational evidence yet proves it—while radiating away their insides over eons of time they become Nature's final entropy producers (Hawking 1974). Eventually—but only very eventually, perhaps on scales of $10^{70}$ years—thermodynamics is the victor, not gravity.

To be sure, approaching gravitational equilibrium (in addition to thermal and chemical uniformity) would mean a Universe full of black holes. At the start of all things—the alpha point of spacetime—perfect equilibrium prevailed for non-gravitational forces, while essentially maximum disequilibrium characterized gravity. As implied by Figure 14, the trends are quite opposite for gravity as for thermal and chemical effects; gravity has counter-thermodynamic tendencies. An initially regular, low-entropy gravitational state is destined to evolve into a highly irregular one at the end of all things—the omega point of spacetime—unavoidably one that maximizes all actual entropies and approximates the "heat death" postulated by classical physics.

All of this is at least developmental change within the phase of stellar evolution, the kind of change that biologist Mayr calls *transformational evolution*—a mostly gradual (and occasionally episodic) alteration among objects quite apart from any generational selectivity. Mountains sprouting in response to tectonic forces, fertilized eggs developing into mature adults, and normal stars swelling into red giants are all examples of transformational evolution. Virtually all changes in the inanimate Universe, among many also in the living world, are minimally of this nature.

But is there any selection going on among the stars—any non-random elimination? In short, is there any *variational evolution* for inanimate objects, as in conventional neo-Darwinism whereby only a minority of objects manage to survive change, in fact only those well adapted to their environments? Perhaps there is, working in consort with other evolutionary drivers, for only stars with sufficiently high values of $\Phi_m$ go on to states of greater complexity; only those main-sequence stars with a free energy rate density above a threshold ($>100$ erg s$^{-1}$ g$^{-1}$) manage to create great order and complexity, as manifest by that state just prior to a titanic supernova event. The Sun, with $\Phi_m = 2$ erg s$^{-1}$ g$^{-1}$ currently, will in about 5 billion years become an ordinary red giant, increasing somewhat its gradient of temperature and chemical composition from core to surface, but it will never fuse carbon into heavier elements, never become more complex than a simple red giant, and never detonate as a supernova. In a manner of speaking, the Sun will not be selected for much greater complexity, as its energy flow, or value of $\Phi_m$, will likely never reach those critical values needed for the natural emergence of greater complexity. To use Mayr's terminology, the Sun will have been non-randomly eliminated from further evolution toward greater complexity.

Much as in the case of biological evolution among living species—a subject soon to be treated—a process of selection, then, also operates in the physical evolution of non-living systems. As will be exemplified in the next section, wherever energy flows are available, living systems develop in ways capable of drawing power competitively and they build whatever structures are needed to optimally process those energies. Selection is a process that tends to reward a system for actions that are reinforced and retained. Similar processes pertain to inanimate objects, even in the absence of genetic replication. Whenever suitable energy flow is present, selection from among many energy-based choices rewards and nurtures those systems that engender pathways capable of drawing and utilizing more power per unit mass—up to a point beyond which too much power can destroy a system. At least as regards energy flow, matter circulation, and structural maintenance while undergoing evolution, stars have much in common with life.

None of this is to claim that stars are alive, a common misinterpretation of such an eclectic analysis. Nor do stars evolve in the strict and limited biological sense; most biologists would say that stars *develop* as part of the larger arena of cosmic development. Even so, the intentionally broadened view of evolution proffered in this book enables close parallels to Darwinian selection, including reproduction liberally judged. Consider the following scenario, reminiscent of Darwin's Malthusian-inspired principle of natural selection: Galactic clouds spawn clusters of stars, only a few of which (the more massive ones unlike the Sun) enable other, subsequent populations of stars to emerge in turn, with each generation's offspring showing slight variations, especially among the heavy elements contained within. Waves of "sequential star formation" (Elmegreen and Lada 1977) propagate through many such clouds like slow-motion chain reactions over eons of time—shocks from the death of old stars triggering the birth of new ones—neither one kind of star displaying a dramatic increase in number nor the process of regeneration ever being perfect. Those massive stars selected by Nature to endure the fires needed to produce heavy elements are in fact the very same stars that often create new populations of stars, thereby both episodically and gradually enriching the interstellar medium with greater elemental complexity on timescales measured in millions of millennia; such rising complexity, further, promotes planetary systems that act as advantageous abodes for life as we know it. As always, the necessary though perhaps not sufficient conditions depend on the environmental circumstances and on the availability of energy flows in such (here, galactic) environments. On and on, the cycle churns; build up, break down, change—a kind of stellar reproduction writ large minus any genes, inheritance, or overt function, for these are the value-added qualities of biological evolution that go well beyond stellar evolution.

Planetary systems can also be analyzed in much the same way as for stellar systems, especially regarding the issue of complexity, again broadly considered. Energy flow, physical evolution, natural selection, and ordered states provide a package of understanding, both for how it is that planets generally become comparable to or a bit more complex than stars, as well as how it was that on (or near) at least one such planet the conditions ripened for the emergence of the even more highly ordered system of life itself. Briefly, the bulk of a planet's order derives from the energy gained via gravitational accretion of raw proto-planetary, initially homogeneous matter, whereupon the onset of energy flow created its geological complexity, from core to surface. In particular, during Earth's formative stage some 4.6 billion years ago when it experienced much of its gross ordering into core, mantle, and crust, its value of $\Phi_m$ was a great deal more then than now. This ought not to surprise us, since almost all of our planet's heating, melting, differentiating, and initial outgassing occurred before the oldest known rocks formed (see again note 5 in Chapter Three). We are referring to the free energy rate density that established thermal and chemical layering within the early, naked Earth (minus an atmosphere and biosphere), the remnant of which geologists strive to explore and model today. Unlike gaseous stars that continue to enhance their thermal and chemical gradients via physical evolution, often for billions of years after their origin, rocky planets order their inhomogeneities mostly in their formative stages while accreting much of their material in less than a million years, after which internal (non-surface) evolutionary events comparatively subside. It is during the early years that planets, at least in their bulk composition, experience the largest flow of energy in their history.

Earth's primordial value of $\Phi_m$ can be approximated by appealing to the conservation of energy (in this case, the first law of thermodynamics applied to a gravitating body), setting the gravitational potential energy of a gas cloud of mass, M, that infalls to form a ball of radius, r, during a time interval, t, equal to the accreted energy gained and partly radiated away while converting that potential energy into kinetic energy, in turn causing a rise in surface temperature, T,

$$\tfrac{1}{2}\,(GM^2\,/\,r) = 4\pi\,r^2\,t\,\sigma T^4\,.$$

Here, $\sigma$ is the Stefan-Boltzmann constant, $5.7 \times 10^{-5}$ erg cm$^{-2}$ s$^{-1}$ K$^{-4}$, the term $\sigma T^4$ being the radiation fluxing through a surface area; the whole right side of this equation equals the total energy budget of the massive blob, namely the product of luminosity and time. The extra $\tfrac{1}{2}$ owes to the so-called virial theorem, which specifies that half of the newly gained energy is immediately radiated away, lest the formative process halt as the heat rises to compete with gravity, thus that part does not participate in formative ordering. The result for early Earth was significant heating, indeed melting, mostly via gravitational accretion, though perhaps aided somewhat later by the decay of radionuclides; however, none of the most abundant radioactive elements, including potassium, uranium, and thorium, have half-lives short enough to have participated in much of this early heat pulse, and thus we neglect them in this admittedly simplified treatment.

In this way, we arrive at an approximate value of $\Phi_m = GM/2rt \simeq 10$ erg s$^{-1}$ g$^{-1}$ for the young Earth, a free energy rate density generally larger than that of the less-ordered Sun yet smaller than that of Earth's more-ordered climasphere, much as might be expected if our philosophy of approach has any validity. With $t \simeq 10^{3-4}$ years, we also find $T \simeq$ 3000 K, a not unreasonable temperature to which ancient Earth might well have been heated during its accretional stage (Lewis and Prinn 1984), to be sure a whole lot less than the approximately 60,000 K to which the assembled rock would have been heated had all the acquired energy been stored internally. The time scale for terminal accretion, that is, the total duration needed to sweep clean the primitive Solar System and to form planets completely, is more like $10^{7-8}$ years, but the solar nebula cooled and mineral grains condensed on the order of $10^4$ years. During this time the bulk of the planets must have formed, or else the loose matter in the solar nebula would have been blown away by the developing "T Tauri" solar wind. Furthermore, slow accretion over the course of millions of years would have allowed the newly gained heat to disperse, resulting in negligible influence on its internal temperatures (of a mere few hundred Kelvins) and thus an inability to melt rock (as opposed to merely heating it), causing minimal geochemical differentiation, if any. By contrast, a temperature of a few thousand Kelvins, as calculated above for more rapid accretion, was surely high enough to have made rock molten, thus helping (along with the long-lived radionuclides) to order our planet's insides as the low-density materials (rich in magnesium and silicon) percolated toward the surface and the high-density materials (rich in iron) sank toward the core, yet not so high as to make this analysis unrealistic. Those same radionuclides and the potential energy realized when huge globs of molten metal plunged into the core would have further heated that core enough to establish a robust magnetic field from the dynamo action of mostly spinning iron. The result is a planet that is today well differentiated, with moderate density and temperature gradients, from core to surface: ~12 g cm$^{-3}$ to 3 g cm$^{-3}$, ~6000 K to 290 K, respectively. We stress that these heating, flowing, and ordering events occurred long ago on Earth; currently, as averaged over our entire planetary globe, the free energy rate density is very much smaller (~$10^{-7}$ erg s$^{-1}$ g$^{-1}$), nor is there much ordering now occurring save perhaps at a few "hot spots" that drive today's surface tectonic activity—and, of course, in the climasphere and biosphere, where much enhanced order is indeed evident, not from energy fluxing outward from inside Earth but from that fluxing inward from outside, indeed from the Sun.

What about the other planets in the Solar System, as well as the debris that never accumulated into the Sun and planets, such as the asteroids and meteoroids? To what extent were these objects ordered in their formative stages long ago, and how do their values of $\Phi_m$ compare to that of early Earth, just calculated? Computations yield a very similar free energy rate density for Venus, much as expected given that its bulk properties (mass, radius, etc.) resemble Earth's closely; Venus is truly Earth's sister planet. Its formation, leading to an ordered core, mantle, and crust (enough to show evidence of volcanism on radar maps), is consistent with a value of $\Phi_m$ (~9 erg s$^{-1}$ g$^{-1}$) virtually equal to that of Earth, ignoring for the moment both planets' atmospheres, where additional complexity has subsequently evolved. By contrast, Mercury and Mars, each having a mass roughly a tenth that of Earth, must have heated and ordered less during their accretion. Their modeled interiors imply only slight differentiation, comparatively smaller cores, and not much global magnetism; indeed their values of $\Phi_m$ (~2 erg s$^{-1}$ g$^{-1}$) are a good deal less than those for early Earth, again as expected for less accreted energy, smaller gradients and flows, and ultimately less complex, now mostly dormant, worlds. Our Moon, too, is not as ordered as Earth and its small value of $\Phi_m$ (~0.5 erg s$^{-1}$ g$^{-1}$) confirms it. Its accretion temperature (~800 K) was sufficient to have heated but not melted much rock or caused much differentiation, consistent with a uniform basalt mantle surrounding a minute metallic core, as inferred from intensive exploration of the Moon in recent decades. So we shouldn't be surprised that our nearest cosmic neighbor is a poorly ordered hunk of rather homogeneous rock; it never did manage to acquire a robust energy flow, from inside or outside, and thus never did become very complex. Asteroidal-sized objects, even as large as Ceres (at ~900 m in diameter, the largest in its class), as well as all smaller objects in our Solar System, experience insignificant accretional energies and very low surface temperatures for any reasonable accretion time scale, thus we should not be surprised to find all of them uniformly constructed and largely unordered.

By contrast with the terrestrial planets, the much larger jovian planets, represented most prominently by Jupiter, are moderately complex. Considering the accretional heating that would have generated the bulk of Jupiter's order, and providing for a somewhat longer accretion time, t $\simeq 10^4$ years, we find a value of $\Phi_m \simeq 30$ erg s$^{-1}$ g$^{-1}$. This is consistent with a priori expectations; just looking at Jupiter would suggest enhanced order and complexity, and its high radiogenic heat flow (even today, equal to twice the energy received from the Sun) would seem to demand it. Not unreasonably stated based on its rather large value of $\Phi_m$, Jupiter's complexity compares favorably to that of a red-giant star. Theoretical models of Jupiter's interior do somewhat resemble the steep thermal and pressure gradients characteristic of a well-evolved star; Jupiter's core is thought to be as high as 40,000 K, its pressure as much as 50 million bars (1 "bar" equaling Earth's atmospheric pressure), both of these a great deal different from its "surface" values of 300 K and 10 bars, respectively. Qualitatively, our comparative analysis seems realistic; quantitatively, despite simplifications, the back-of-the-envelope energy flow calculations work out well. Thus, one can practice comparative planetology regarding order, flow, and complexity by undertaking the same kind of thermodynamic analysis performed above for stars, as well as for a wide spectrum of ordered systems encountered at the end of Chapter Three.

Our discussion of physical evolution is not yet done. We seek a graceful segue to the upcoming commentary—unapologetically thermodynamic commentary—on biological and cultural evolution in the final sections of this Discussion, which brings us back to Earth once more. When ancient Earth was forming and undergoing the bulk of its physical evolution several billion years ago, its value of $\Phi_m$ was then of order tenish erg s$^{-1}$ g$^{-1}$. By contrast, that value now is merely a fraction, currently less than a millionth, of its formative value, implying, as expected, that Earth in bulk—its geological globe—is now *relatively* dormant, at least compared to its more violent past. Earth's ordering, in the main, is over; its structural complexity developed long ago when vast quantities of accretional energy flowed through our primitive planet. Not that Earth is now geologically dead, far from it; close to the surface, plate tectonics continues to remold the land and sea scapes, albeit at glacial pace. At some specific locales on Earth, evolutionary activity is much in evidence, and significantly larger values of $\Phi_m$ prove it. Despite the cooled, differentiated, and long-since ordered Earth, a variety of concentrated sites exhibit enhanced energy flow and consequent increased ordering and construction even today: steam geysers, active volcanoes, and oceanic trenches are all examples of Nature's geological heat engines, most of them interacting with the biosphere. Even earthquakes, violent and sudden forces that tend to upset our lives and destroy our cultural edifices, can be considered sources of renewed geological ordering; those forces are basically driven by remnant energy left over from Earth's formation, now distributed in its hot core and in mantle convective cycles (à la Bénard cells) that power continental drift. Of all these geothermal sources of energy today, the underwater, mid-ocean ridges are perhaps most intriguing and instructive, for it is in the upwelling waters of the hydrothermal vents that geosystem organization is most discernable. Furthermore, rich and complicated ecosystems of numerous and diverse life forms are powered by suboceanic heat engines, the vents themselves displaying much localized enhancement of order, flow, and complexity on Earth.

Also known as "black smokers" (owing to sulfur-laden emission), submarine vents are narrow seafloor crevices through which pressurized hot water is ejected from subterranean regions; substantial temperature gradients and thus much heat throughput are characteristic features. Cold seawater having a temperature of $\sim$275 K interacts with much hotter water rich in metal sulfides leached from molten rocks below, in fact superheated to $\sim$600 K at the vent orifice, the whole mixture upwelling at $\sim$250 bars of pressure through vertical chimneys in the suboceanic crust. Such hot, rising fluid is the source of the energy flow that drives and sustains much biological activity in the vents, but not conventional life forms familiar to us, like most plants and animals whose cells have true nuclei that divide by mitosis. For example, autotrophic methanogens (simple procaryotic bacteria) thrive at the scalding temperature of the mixture near the central conduit of the chimney, requiring >350 K for optimum growth but not much higher than 385 K. Sometimes called "extremophiles" or "hyperthermophiles" given their high-temperature requirements, methanogens differ from common life forms elsewhere on Earth that make a decent living at or near the surface, where environmental conditions seem to be homeostatically controlled at roughly "room temperature," ~290 K (Lovelock 1979). They differ in another way as well: Despite their autotrophic traits, they are not dependent on $O_2$-producing photosynthesis and a visible-light energy supply. No sunlight reaches these submerged vents. Instead, subterranean methanogens operate a parallel process known as chemosynthesis, drawing their (chemical) energy from the oxidation of hydrogen sulfide ($H_2S$) within a dark and deep biosphere wholly different from the familiar one near Earth's surface. These are among the so-called archaebacteria, a category of life that displays much diversity and durability at the base of a food chain and that symbiotically lives within a remarkable community of two-meter-long tube worms, ten-kilogram giant clams, and idiosyncratic microbes thriving under what we at the surface would call extraordinarily harsh conditions.

Undersea hydrothermal vents are far removed from thermodynamic equilibrium; their steep temperature gradients guarantee it. Although the measured heat flux for the whole Earth averages 63 erg cm$^{-2}$ s$^{-1}$, flux values often reach nearly an order of magnitude higher at the mouth of the vents. And since the typical vent of cross-section ~1 m$^2$ and depth ~100 m is so small (compared to all of Earth), the enhanced heat renders vastly larger free energy rate densities than the mere $10^{-7}$ erg s$^{-1}$ g$^{-1}$ flowing, on average, through our planet today. Values of $\Phi_m$ for such vents can reach of order unity, occasionally somewhat higher. Not surprisingly, with these kinds of striking non-equilibrium energy flows, undersea vents are among the premier examples of localized growth of order and complexity within geologically open systems on Earth today. But truly surprisingly (at least when first explored in situ by the submersible vessel *Alvin* two decades ago; Corliss et al., 1981), the vents appear to be more than a mere source of geological order. They could well be the engine that drove the early emergence of biological order as well, an issue not without controversy for those who prefer that life began on or near Earth's surface, or perhaps even fell from the sky already assembled and embedded within rocks or comets travelling through interplanetary space ("panspermia").

Given that the archaebacteria—representing most of the entries, along with the eubacteria and eukarya, in the newly revised, three-domain bush of life (Woese 1994)—are probably the most ancient life forms, or at least close to life's last common ancestor (harking back to the Archaea period in the early Pre-Cambrian), the hydrothermal vents have become a leading candidate for suitable sites where life on Earth originated several billion years ago. The contention of some researchers (Shock 1990) that our planet's surface would seem to have been inhospitable for life's origin—possibly lacking a reducing environment conducive to life, probably lacking protection from harsh solar ultraviolet radiation and incoming asteroids once life formed, and almost certainly lacking sufficient ammonia to help form life—bolsters the idea that deep-sea vents might have provided environmental conditions better suited for the origin of life. RNA sequencing further implies that the earliest organisms were thermophilic, allowing survival in oceans that were heated by volcanoes, hot springs, and asteroidal impacts. Accordingly, the vents of the underlying crust of prebiotic Earth—reducing, protective, and rich in ammonia—could have acted as hydrothermal reactors that literally fashioned biology out of a geological setting. With the sole energy source being geothermal (and the Sun playing no role), the free energy rate density can be approximated by appealing to an enhanced flux of $10^3$ erg cm$^{-2}$ s$^{-1}$ through the 1-m$^3$ volume of a typical flow reactor within an otherwise larger vent chimney, where prebiotic species concentrate and interact in water of density 0.2 g cm$^{-3}$ at 600 K. The answer is $\Phi_m \simeq 50$ erg s$^{-1}$ g$^{-1}$, a value quite uncertain given the many unknowns among vent properties, yet well within the expected complexity range between the primordial Earth and cellular photosynthesis.

Planets, then, like stars before them, do seem to display a trend in complexity, at least as characterized by the flow of free energy density through open, non-equilibrium systems. Jupiter is more complex than Earth, Earth in turn more complex than Mars, Mars more than the Moon, and so on. It would seem, from study of our Solar System alone, that mass is a deciding factor: the more massive, the more complex. But this clearly cannot be the bottom line, generally, for all structured systems; complexity cannot be solely a function of mass. Normal stars and typical galaxies, hugely more massive than any planet, are less ordered, less complex. And plants, animals, and brains, much less massive than the planets, are more orderly and more complex. Why? Because of selectivity. These latter, truly complex systems were favored by Nature; they were naturally selected.

## Evolution: A Biological Perspective

Early, promordial Earth was surely too hot for amino acids and nucleotide bases to survive; there were no hiding places then—no cool lithosphere, no dark hydrosphere, no protective atmosphere. Remarkably, however, as soon as our planet cooled enough for rocks to form, energy sources to diminish, and the atmosphere to thicken, life arose. Geological dating by radioactive means and paleontological observations of deposited strata reveal that the oldest rocks solidified ~4 billion years ago, cellular life appeared ~3.8 billion years ago, and life began to express its major metabolic features as long ago as ~3.5 billion years (Schopf 1999). If life did emerge indigenously on primitive Earth (as opposed to arriving intact from space)—and almost all working scientists agree that it probably did—then it must have done so as soon as it was chemically possible. The gap between the oldest fossils and the oldest rocks grows smaller with each new discovery, the implication being that the pre-cellular phase of chemical evolution must have been astoundingly rich in invention and innovation, much as biological evolution has subsequently produced a cornucopia in number and diversity of species in more recent times. The difference is that biological evolution leaves a record preserved in the rocks, whereas chemical evolution does not.

The Introduction to this book encapsulates how physical phenomena and environmental conditions can tease order from chaos, and not merely regarding inanimate objects. Impressive amounts of order emerge in laboratory experiments that simulate primeval Earth and the production of life's building blocks, often with the aid of agents that constrain chance. Catalysts, for example, play prime roles in governing the rates at which molecules react, thereby removing some of the randomness, either in the test-tube simulations of life's origin or among today's life forms. Fine-grained mineral clays, with their crystalline structures and "lock-and-key" surface geometries, could well have acted catalytically on early Earth, collecting certain small molecules and not others, thereby guiding them to interact and connect into larger ones. There is little mystery here, as the clays' ions preferentially attract matter around them to react in specific, non-random ways; clays also act as a desiccator, removing water to allow larger molecules to assemble from smaller fragments. Such selectivity clearly operated in the realm of chemical (i.e., pre-biological) evolution, an integral part of the loftier cosmic-evolutionary scenario.

In that hazy interface between life and non-life, a kind of natural selection—to distinguish it, let's call it chemical selection—was likely as important then, at the dawn of life, as in biological evolution now. Chemical selection doubtlessly rewarded reaction pathways and cycles that remake moderately sized molecules needed in the further synthesis of increasingly complex organic products, thus speeding reactions that would otherwise proceed sluggishly in a dilute, pre-life environment. Such so-called autocatalytic cycles, acting like positive feedback loops, and especially integrated sets of these cycles, termed "hypercycles," can increase catalytic efficiency and specificity in the production of key polynucleotide and polypeptide sequences. In this way, chemical selection could have regulated reaction pathways, choosing some while rejecting others, thus making molecules capable of cyclical replication even before a primitive metabolism arose. This was Schrödinger's (1944) predisposition: biological replication based on his beloved quantum mechanics could have existed without, and therefore preceded, thermodynamically inspired cellular metabolism. Today, we would surmise that if one product among a mixture of possible products displays autocatalysis, then the chemical reaction producing it is favored over other possible courses of reaction. Pre-Darwinian selection would have provided a dose of determinism, which, when working alongside chance, might well have created molecules resembling (or equal to) the guanine, cytosine, adenine, and uracil bases comprising RNA today—molecules capable of replication thanks to base pairing, whereas molecules with other bases unable to engage in base pairing were weeded out. Not far behind would have been the replicating DNA ancestors of all living things, again the result of Darwinism among the molecules. If so, then the roots of life indeed penetrated the realm of non-living matter. These speculations are based mostly on mathematical models of feasible chemical routes available to primitive molecules on early Earth, and are favored by theorists who argue that "selection does not work blindly . . . [rather] it is highly active, driven by an internal feedback mechanism that searches in a very discriminating manner for the best route to optimal performance" (Eigen 1992).

By contrast, empiricists, informed more by laboratory experiments, argue that the emergence of metabolism more likely preceded replication. And in modern organisms, metabolism (comprising cell structure, boundary membrane, and enzymic activity) is strictly dependent upon the existence of proteins. Actually, it is not only some experimentalists who favor metabolism first. The physicist Freeman Dyson (1985) has proposed a theoretical model for two origins of life, once with cells and later with genes, stressing diversity and error-tolerance as life's salient characteristics. Yet it is the experimentalists, mostly, who maintain that semi-permeable membranes—the border between cells themselves and their immediate tissue environment governing the two-way transport of small molecules—must have been an essential prerequisite for any early replicating molecules, lest they have no stable locale, or discrete unity, for the concentration of a cell's chemical components. Laboratory experiments that go beyond the simple Miller-type syntheses reviewed in the Introduction demonstrate that, with energy easily accessible on primitive Earth, amino acids concentrate further into 1–3μm-diameter "proteinoid microspheres," which are essentially organized, organic condensates displaying a remarkable kind of cell-like metabolism—and at least a mechanical form of replication (S. Fox 1988). One of the recurrent criticisms of these advanced experiments, however, is that such polymerization requires the initial amino-acid mixture to be heated to a temperature above the boiling point of water, but now that geothermal vents have become leading environmental contenders for the onset of life, this liability might be an asset. Although these condensates are not true proteins in the sense of having biological activity, it would seem that Nature is playing a malicious joke on us if they are not examples of organic clusters somewhere on the chemical-evolutionary road to life. The question remains: Which came first, naked genes or protobionts?

Considered by some a procedural compromise yet by others an experimental breakthrough, an "RNA world" on early Earth seemingly solves this chicken-or-egg dilemma by asserting that primitive ribonucleic acid acted as both replicator *and* catalyst; such a "ribozyme" would have performed double duty by storing genetic information and catalyzing its own replication. RNA molecules that can catalyze biochemical reactions are known to exist (Cech 1990), and the molecular evolution of these ribozymes could have been subject to selective pressure, like the ability to hydrolyze compounds faster, given variations among early RNA sequences on Earth billions of years ago (Gesteland et al. 1999). Eventually, somehow, that RNA world collapsed or evolved into one obeying the central dogma of modern molecular biology: DNA $\leftrightarrow$ RNA $\rightarrow$ protein (the arrows signifying transfer of genetic information, not chemical reactions), the last of which often act as catalysts in today's living systems. Whether ribozymes self-polymerized the first long nucleotides and eventually life itself, or enzymic proteins emerged alone (perhaps in hydrothermal sites), or even whether clays or other environmental catalysts acted as templates for life, remains unknown (Shapiro 1986). The origin of life, along with the origin of galaxies, represent the two chief missing links in all of cosmic evolution.

We return now to considerations of thermodynamics, as we almost always do; that is our convergence in this book, admittedly our bias. Energy flow was an inevitable result of the non-equilibrium conditions needed for either route toward life, the central role of which has been clear to biophysicists for decades (Morowitz 1968). The source of that energy supplied to any pre-life system could have taken the form of solar radiation, atmospheric lightning, shock waves, energy-rich chemicals, geothermal heat, or any combination of these or other suitable sources undoubtedly available at the time. Mostly likely, gradients and flows would have pooled resources to organize both replicating and membrane molecules among effusing protocells, after which chemical selection presumably favored autocatalytic cycles capable of making products that can capture and store information in macromolecules. Here, then, is also the beginnings of ecology, which seeks to place life into the general context of energetics.

The chemical evolutionary steps that led to the origin of cells follow, in the main and in principle, a straightforward sequence: acids and bases → proteinoids and polynucleotides → protocells and life. In practice in the lab, this sequence does indeed seem to occur, and rapidly and simply at that, each of the open systems along the path consuming more energy and thus growing in complexity. Provided the energy is nurturing, the ingredients appropriate, and the environment protective, the organic molecules increasingly self-assemble, or self-organize, into cell-like blobs, much as hypothesized for the origin of the first cell by biologist George Wald (1954) in a classic paper of nearly a half-century ago. All of which brings to mind another pioneer, Louis Pasteur, who posed his famous, lingering question during a theatrical lecture in 1864 at the Sorbonne, where he disproved the popular nineteenth-century doctrine of spontaneous generation as viable for life's origin: "Can matter organize itself?" As for galaxies, stars, and planets, the answer for life forms is apparently in the affirmative—self-assembly, yes, but not without energy flowing.

The energy of synthesis for a few representative biomolecular building blocks will further our case, each of them members of the group of twenty amino acids comprising the protein machinery that executes essentially all the functions involved in life as we know it—storing and transferring matter, energy, and charge, performing catalysis, controlling reactions, and acting as the DNA-determined and covalently linked substances of all organisms. Glycine ($CH_2NH_2COOH$), with a molecular weight of 75 amu ($1.2 \times 10^{-22}$ g) and the simplest side chain of a single hydrogen atom, requires some 370 kJ mole$^{-1}$; lysine, a somewhat more complex amino acid with a weight of 146 amu and a linear side chain of one nitrogen and four carbon atoms, requires 1530 kJ mole$^{-1}$; and tryptophan, the most elaborate of these acids with a weight of 204 amu and a multiple-ringed, carbon-nitrogen side chain, requires 2390 kJ mole$^{-1}$. Not surprisingly, the energy needed to fashion even these simple molecules is proportional to the complexity of their being.

Controversial though they may be, the proteinoid microspheres noted above would have been among the most elementary life-like systems. These alleged protocells contain no recognizable proteins as such, but they do harbor collections of myriad protein-like polypeptides, and evidently utilize energy flows in accord with their degree of complexity. For example, a rough estimate of free energy rate density—our postulated universal measure of complexity—for a 2μm-diameter (*E. coli*-sized) microsphere harboring typically $10^{-11}$ g of synthesized organic polymers yields $\Phi_m \simeq 200$ erg s$^{-1}$ g$^{-1}$. This value assumes an energy flux of 400 erg s$^{-1}$ cm$^{-2}$, which is typical of that for undersea geothermal vents, although high for a single atmospheric electrical discharge or for solar ultraviolet radiation reaching Earth's surface (shortward of the 2500 Å wavelength needed to break and reform, for example, O–H, C–H, or C=C bonds of about $10^{-11}$ ergs (~5eV) each, and corrected for one-third less solar luminosity in primordial times—see note 2). Admittedly, the energy input is not well known as we are still uncertain what event really did trigger the origin of life. It is unlikely that any single source of energy could account for all the organic compounds on primitive Earth, but this broadly representative value of $\Phi_m$ does put it in a reasonable range midway between the complexity of embryonic Earth ($\Phi_m$ of order 10 erg s$^{-1}$ g$^{-1}$) and that of photosynthesizing plants (roughly 900 erg s$^{-1}$ g$^{-1}$). We are crossing the interface of chemistry and biology, indeed of astronomy and biology, and our proposed synthesis, broadly construed and predicated on energy flow, seems to be holding up.

Energy-flow diagnostics show a disconcertingly wide range of $\Phi_m$ values when analyzing contemporary unicells, the smallest and simplest entities (save viruses) that everyone would agree are definitely alive. We seek to characterize how energy is used for synthesis and transport of molecules, for performance of functional work, and for reserve storage in energy-rich molecules, among other tasks. Surprisingly, a hugely diverse array of single-cell types and actions confront us, making even back-of-the-envelope calculations tricky; sizes and shapes of single cells vary greatly, as do their metabolic and reproductive rates, for there is no idealized, textbook cell. Some bacterial cells (such as *pneumococcus*) are as small as 0.2μm in diameter and $10^{-15}$ g in mass, compared to typical liver cells that measure a hundred times as large and nearly a million times as massive; nerve cells can be longer than a meter and appropriately more massive still. Metabolic rates for cells can also span many orders of magnitude, making consequent values of $\Phi_m$ cover a spectrum so wide as to make it difficult to find a typical value for a typical, modern cell, let alone a primitive one that might have eked out a living on early Earth. *E. coli,* the 2μm-diameter microorganism about which perhaps more is known than any other unicell, can replicate as quickly as 3 times per hour under maximally robust (37.5°C, or 310.5 K) conditions, such as a delicious nutrient broth in a warm petri dish or the cozy gut of a healthy human; its metabolic rate can be approximated knowing that laboratory studies show 0.015 erg of heat associated with the biosynthesis of each microorganism (2 $\times 10^{-12}$ g) of *E. coli,* so that when normalized to its peak 22-minute reproductive cycle we find $\Phi_m \simeq 10^6$ erg s$^{-1}$ g$^{-1}$. This prodigious value for the free energy rate density (probably a survival-related feature since bacterial cells live in environments over which they have little control and from which they cannot escape) is, however, tempered by the knowledge that most individual cells do not operate under such peak conditions nor do they have such fearsome reproductive rates, and for these the metabolic rate is less; for instance, $O_2$-consumption measures show *Paramecium* to have $\Phi_m \simeq 10^4$ erg s$^{-1}$ g$^{-1}$, heat transfer during stimulation show nerve cells to have $\Phi_m \simeq 40$ erg s$^{-1}$ g$^{-1}$, and a whole suite of bacteria, algae, and fungi that sluggishly metabolize in pore spaces within barren rocks in the otherwise lifeless, extremely dry valleys of Antarctica have values of $\Phi_m$ close to unity (Lehninger 1971). We conclude, despite the extreme diversity of cell types, that the average value of cellular free energy rate density (hundreds to thousands of erg s$^{-1}$ g$^{-1}$) does agree reasonably well with that expected at the interface of non-life and life—but we shall revisit at the end of this section these impressive little microbes.

Onward across the bush of life—cells, tissues, organs, and organisms—we find much the same story unfolding. Studies of life's precursor molecules, as with life itself, show the same *general* trend found earlier for stars and planets: The greater the perceived complexity of the system, the greater the flow of free energy density through that system—either to build it, or to maintain it, or both. Energy, for sure, is essential to every aspect of life. But not just total energy incident on living systems; a more detailed study of life would distinguish the inputs containing more concentrated energy from the dispersed wastes, for it is this energy difference that is used by organisms to order and maneuver the atoms in their bodies. Origin, differentiation, growth, and evolution all involve energy-requiring syntheses of vital substances, such as nucleic acids and proteins. Mechanical work done by muscles and limbs, electrical impulses in nerves and brains, active transport of substances against osmotic concentration gradients, and temperature-maintenance of warm-blooded animals, to name but a few examples of biodynamics, all reduce to flows of energy through open systems. That energy is generated by the oxidation of foods consumed by the body and subsequently supplied to the tissues in the form of chemical energy released by metabolic reactions of specific "energy-rich" compounds. These are the ATP molecules discussed, however briefly, at the end of the Introduction, therein to check that all accords with the second law of thermodynamics, and herein to underscore the mechanical, chemical, electrical, and thermal work upon which all life depends.

This book is certainly not the first to stress the role of energy in constructive matters, although most others have been either qualitative in approach or restrictive in application. In particular and as an example of the latter, Morowitz (1968) has pioneered the study of energy flow in biological systems, not only showing how such non-equilibrium (or as he calls them "steady") states can acquire properties very different from those that might be expected in equilibrium, but also asserting that biological evolution, function, and increased organization can be studied as a physical process. He furthermore goes on to say that "from the study of energy flow in a number of simple model systems, . . . the evolution of molecular order follows from known principles of present-day physics and does not require the introduction of new laws"—a sentiment roundly embraced here, for systems both alive and abiotic.

What about higher-order living systems, especially their attendant changes (both ontogenetic and phylogenetic) in biological organization? Complexity must increase at life's origin and, again *generally,* during its development and evolution, for living systems are demonstrable storehouses of concentrated energy, low entropy, and much order. After all, intricate organisms, whose bodies contain many different organs each with specialized structures and functions, are surely more ordered and complex than one-celled creatures. As with other objects in the Universe, the information and complexity sciences, alongside a good dose of thermodynamics, can be used to describe the structural (if perhaps less so the functional) aspects of biological organization, much as for highly evolved stars wherein more information is needed for their description. Konrad Lorenz (1977) summed up our agenda decades ago, proposing (though again only in words) that energy transfer and information processing work in tandem to grant advantages to individuals operating among the selective pressures of the biotic environment; together, they account for the mode and tempo of biological evolution: "Life is an eminently active enterprise aimed at acquiring both a fund of energy and a stock of knowledge, the possession of one being instrumental to the acquisition of the other. The immense effectiveness of these two feedback cycles, coupled in multiplying interaction, is the pre-condition, indeed the explanation, for the fact that life had the power to assert itself against the superior strength of the pitiless inorganic world."

Biological systems that are clearly living enjoy value-added order compared to physical systems that are clearly not. The former are not only more ordered structurally, but they also display functionality and adaptation. Whereas structural complexity in biology is largely related to covalent bonding, functional complexity is more a case of metabolic cycling of materials—inflowing, regurgitating, and outflowing of nutrients essential to life. Functionally, organisms are active entities, constantly doing something in order to maintain their order in the face of potentially disruptive environmental forces. Living systems *actively* exist, as opposed to mere physical systems like an inorganic crystal or a dormant planet that more passively comprise an ordered state once formed. Both physical and biological objects possess unadulterated structure, yet living systems need that structure organized regularly (by means of energy flows) in order to perform functions, such as lungs breathing or hearts beating, which are vital to the continuance of organisms' existence. Functional properties are therefore sophisticated add-ons to life's structural properties, representing an additional degree of refined functional order superposed on its more conventional structural order. Perhaps that is why some, like Dobzhansky et al. (1977), assert that function is the most important concept in biology, enhancing life (lest it be merely a phenomenon of lowish-temperature physics) above and beyond inanimate complex systems. That may also be why others, like Wicken (1987), claim that empirical measures of function are "a very perilous enterprise [that] we aren't even close to knowing how to quantify."

Genetic inheritance is a second factor granting living systems enhanced order over and above that of physical systems. Of course, these two value-added features of life are related since the genetic code *is* functional information; nucleotides of DNA have the specific function to direct amino acids into protein structures. Here adaptation is center-stage, a pivotal feature of neo-Darwinism—descent with modification via reproduction—and yet another source of order that only a huge stretch of the imagination could grant to physical systems alone. Each new living descendant begins its ontogenetic growth not from scratch, but rather by inheriting many of its character traits from its ancestors, which in turn inherited many of theirs from their ancestors, and so on back in time. Consequently, newly arising, advantageous traits can accumulate in a population over time, thereby becoming fixed, inherited properties that add to any extant order. In addition to order generated at the start of organisms' lives, then, phylogenetic change and speciation can grant further order, at least for those life forms able to adapt successfully and survive. Neo-Darwinian selection always tends to generate biological order because it increases the proportion of relatively fit variants while decreasing their unfit counterparts. The story of this enhanced order is the long historical narrative of rich though ruthless biological evolution so grandly played out on planet Earth for billions of years (Fortey 1997).

That said, it seems doubtful that a strict adaptationist program can explain all order, form, and complexity throughout the whole of the living world. Those mechanisms (not yet all understood) governing phylogenetic change presumably derive in part from energy flows, and thus to the dramatic rise in the free energy rate density, $\Phi_m$, in recent times. That is, although the genome surely specifies the many varied traits of life, energy—some would say information, but I prefer energy—is still needed to help realize those traits. Life forms displaying stasis—the cessation of change for long durations—are not merely well-adapted to a stable environment but are also the recipient of optimal energy flows from and to that environment. Natural selection itself would seem to be, at least in part and as with Lotka (1922), a process favoring organisms whose energy-capturing sensors more efficiently utilize free energy, thereby aiding the preservation of species. Should that incoming energy (rate per unit mass) exceed some threshold, thereby driving the system further from equilibrium (see Figure 9), then entirely new species capable of effectively handling the enlarged flow of energy may be selected, again quite naturally. The Canadian ecologist Lionel Johnson (1988) captured some of these ideas when discussing the thermodynamic origin of ecosystems: "Evolution is the outcome of the ultimate ascendency of the trend toward increasing diversity and acceleration of the energy flow, counteracted and retarded by the individual species attempting to proceed in the direction of greater homogeneity and deceleration of the energy flow."

Can the generally increased complexification of evolved life forms be explained partly by the operation of physical laws and irreversible events on the macroevolutionary level, in addition to the conventional accumulation of small, inherently reversible microevolutionary changes in populations? Such a proposition would have the order won by neo-Darwinism superposed—literally as value added—on the order promoted by energy flows so prevalent among inanimately ordered systems. Surely, at a fundamental level, biological systems are specifically configured physical systems, subject to all the thermodynamic properties and constraints discussed earlier. Besides its importance for biological structure and function, energy flow through dynamical, dissipative living systems might well be basic to the understanding of biological evolution as well; neo-Darwinism does seem to improve the management of energy flow with each significant increase in organismal complexity. Many groups of plants and animals first appear in the fossil record as large, simple structures, later to be replaced by smaller, presumably more efficient species; armored fishes, giant ferns, huge trees, expansive birds, and dinosaurs gave way to smaller species, for miniaturization provides more functionality for less power and less mass, ultimately raising values of $\Phi_m$ and therefore of complexity with evolutionary success. Although smacking of reductionism, given their physical-science posture, these statements nonetheless embrace the holism typified by the functionality and adaptationism just noted—much in keeping with our avowed premise in the Preface that the cosmic-evolutionary agenda is by human design a synthesis of these two great philosophies of approach: Reductionism and holism need not be incompatible. Others have been down this track before with like-minded ideas, most notably the bold and ambitious thesis ("biological evolution is an entropic process") of Brooks and Wiley (1988), who have attempted a unification of basic biology with basic physics, although they eschewed energy flows in favor of information capacity and content in order to describe the observed hierarchy of complexity in organic evolution on planet Earth.

What about survival, that quintessential Darwinian issue often raised but seldom understood? Just who are the fittest? Energy flow might well confer advantage in the so-called struggle to survive; complexity itself might promote survival—for some. When expressed in basic terms such as the food gathering needed for life, it would surely seem that energy intake must be survival-related. With complexity comes variety and biodiversity, thereby enhancing opportunities for organisms to increase their chances for survival and reproduction. Yet with complexity comes specialization and vulnerability as well, for each part of a complex system has a particular job it must do, implying that viability, or fitness, might inversely correlate with complexity. For example, in biology, the survival of a mammal depends on the successful powering, coordinating, and functioning of many highly sensitive organs, each one of which must work individually and in tandem; a koala bear's exclusive dependence on eucalyptus supply represents not only a pinnacle of efficiency and adaptation for that life form but also a serious problem should its environment change (Lovejoy 1981). Likewise in culture, complex machines are often inherently less reliable than simple ones, implying that complexity and its inevitable specialization might well carry disadvantages; compare the reliability of a multi-media presentation using a high-end laptop with that requiring a simple set of 35-mm slides. One of the foremost paleontologists of a half-century ago, George Gaylord Simpson (1944), often argued that it was the generalists who have the upper hand: "Survival of the relatively unspecialized." The notion that complex systems are more prone to extinction does not, however, preclude an evolutionary drive toward complexity, recognizing that, although many will be engaged, few will be selected to travel new routes toward robust energy rate densities; genetic mutations (that's chance) occasionally produce organisms a bit more efficient than other members of a species in utilizing available energy (that's determinism), thus making those organisms better able to survive.

All things considered, biological systems are best characterized by their coherent behavior, for their maintenance of order requires a great number of metabolizing and synthesizing chemical reactions as well as a host of elaborate mechanisms controlling the rate and timing of many varied events. Furthermore, such living systems evolved in the past within environments rich in energy flux, and thus have inherited the means to acquire the needed energy flow via metabolic processes. The pathways open to biological evolution are constrained, not because few solutions exist but because energy resources are limited; natural selection exploits energy flows, determining which flows are conducive to the system, thereby apparently optimizing them. Admittedly, the role of energy flow is more clear for growth, development, and metabolism than it is for reproduction, population differentiation, and speciation—the entropic nature of biological evolution at the functional level less obvious than at the structural level. Yet none of this means that life violates the second law of thermodynamics, a popular misconception. Although living organisms manage to decrease entropy locally, they do so at the expense of their global environment—in short, by increasing the overall entropy of the remaining Universe, as demonstrated at the end of the Introduction for the case of any organism tending to the needs of its daily metabolism, or in like manner the case of any ordered star forming over eons from an interstellar cloud.

Much as for massive stars, planetary climaspheres, and other ordered systems, living things, too, are able to circumvent locally and temporarily the normal entropy process by absorbing available (free) energy from their surroundings. Not at all in disagreement with thermodynamics, it is a remarkable testimony to any system's adaptability that a mere rearrangement of its components allows, within a certain space and for a limited time, the growth of order *in accord with the second law.* Life does so during its formation, development, and evolution owing to temperature gradients throughout Earth's diverse environment. The gradient is essential and so is the environment; life itself has no such gradient, for all organisms are essentially isothermal. What is the origin, then, of the thermal differences and ultimately of the energy utilized in the process of living? On Earth, it is our Sun. Energy flows from the hot ($\sim$5800 K) surface of the Sun to our relatively cool ($\sim$290 K) planet. Useful work can be done with this available energy, and the capture of direct sunlight by photosynthesizing plants to convert $H_2O$ and $CO_2$ into nourishing carbohydrates is a premier example; animals also obtain the Sun's free energy indirectly by eating other life forms. This, again, is the decades-old Schrödinger argument that an organism "keeps aloof from the dangerous state of maximum entropy by continually drawing from its environment negative entropy." Such colorful phrases that go on to assert that life "continually sucks orderliness from its environment" are tantamount to saying that heterotrophs and autotrophs import low-entropy molecules and high-grade photons, respectively, to run their metabolisms, after which low-grade photons are radiated away and high-entropy molecules excreted. In effect and simply stated, what Schrödinger meant is that any organism's existence depends upon increasing the entropy of the rest of the Universe.

Of course, if left alone, all living things, much like everything else in Nature, tend toward equilibrium. Just twitching a finger, batting an eyelid, or merely thinking while reading this book expend some energy. Any action taken indefinitely, without further energizing, would drive us toward an equilibrium state of total chaos or orderlessness. By contrast, living systems stay alive by steadily maintaining themselves far from equilibrium. They do so by means of a flow of energy through their bodies. In point of fact, unachieved equilibrium can be taken as an essential premise, even part of an operational definition, of all life, as indeed we did at the start of Chapter Three.

Human beings, for example, maintain a reasonably comfortable steady-state by feeding off our surrounding energy sources, principally plants and other animals. We say "steady-state" since, as argued earlier for any open system, by regulating the rate of incoming energy and outgoing waste, humans can achieve a kind of stability—at least in the sense that while alive, we remain out of equilibrium by a roughly constant amount. In a paradoxical juxtaposition of terms, we might therefore describe ourselves as "dynamic steady-states," a phrase already used in the Introduction to characterize dissipative systems—of which we are one such example among many. Nor are we perfect or even efficient, for we squander much of the acquired energy while radiating heat into the environment; warm-blooded life forms are generally warmer than the surrounding air, humans in particular having the radiative power (in the infrared) of about a 100-watt light bulb. Even so, some of the absorbed energy can drive useful work, thus keeping our wonderful human edifice (and it is wonderful!) alive and metabolizing. Once this available energy flow ceases, the dynamic steady-state is abandoned and we drift toward the more common "static" steady-state known as death, where, following complete decay, the elements of our bodies reach a true equilibrium.

Here is what happens in the food chain consisting of grass, grasshoppers, frogs, trout, and humans. According to the second law, some available energy is transformed into unavailable energy at each stage of the food chain, thus causing greater disorder in the environment. At each step of the process, when the grasshopper eats the grass, the frog eats the grasshopper, the trout eats the frog, and so on, useful energy is lost. The numbers of each species required for the next higher species to continue decreasing entropy are staggering. The support of one human for a year requires some 300 trout. These trout, in turn, must consume about 90,000 frogs, which yet in turn devour approximately 27 million grasshoppers, which live off some thousand tons of grass. Thus, for a single human being to remain "ordered" (i.e., to live) over the course of a single year, we need the energy equivalent of tens of millions of grasshoppers or a thousand tons of grass. Clearly, then, we maintain order in our bodies at the expense of an increasingly disordered environment, which nonetheless is no reason to ravage it unnecessarily. The only reason that Earth's environment does not decay to an equilibrium state is the daily dose of sunshine, and that, in turn, owes to cosmic expansion; if the Universe had not broken its primeval radiation-matter symmetry some 100 millennia after the big bang, there would be no stars, no planets, no us. The whole biosphere comprises a non-equilibrium state subject to regular and vibrant solar heating, thus reducing much entropy locally. Earth's thin outer skin is thereby enriched, permitting us and other organisms to go about the business of living, exploring, and wondering.

It is instructive to pursue this point a bit further, for it sometimes seems preposterous that life depends, ultimately, on the expansion of the Universe. If Earth's atmosphere and outer space were to achieve thermal equilibrium, energy flow into and out of Earth would cease, causing all thermodynamic processes on our planet to decay within surprisingly short periods of time. A rough estimate shows that the reservoir of Earth's oceanic thermal energy would become depleted within a few months, the latent heat bound in our planet's atmosphere would dissipate in a couple of weeks, and any mechanical energy (such as atmospheric circulation that contributes to weather as we know it) would be damped in a few days. So be sure to place Earth's energy budget into perspective; neither our planet's primary source of energy nor its ultimate sink is located on Earth.

Furthermore, take heed and use caution while regarding evolution as progress. Even as humankind goes about its daily routine, as impressively ordered as civilization is, disorder rises in the Earth-Sun environment. The principal source of life's free energy, the Sun, is itself running down as it "pollutes" interstellar space with increasing entropy. Evolution means creation of ever more complex islands of order at the expense of even greater seas of disorder elsewhere in the Solar System, as well as the Universe beyond.

Let us mine this lode deeper while simultaneously making contact with quantitative estimates made earlier. Not only is life, at any given moment, a reservoir of order, but biological evolution itself also seems to foster greater amounts of order from disorder. Whether it be mostly by precise molecular replication of genetic material or by random errors in replication—necessity or chance again, and usually the marriage of both—neo-Darwinism is a genuinely creative process, giving rise to novel, mostly unpredictable, forms of order. During the course of biological evolution, succeeding species in a given lineage often (but not always) become more complex and thus better equipped to capture and utilize available energy; the incoming energy, if suitable in quantity and quality, contributes to the success of the species and thus the growth of order, the whole process replete with feedback loops that reinforce both energy flow and evolutionary complexity. Moreover, as stressed by Mayr (1982) well before Eldredge and Gould coined the term "punctuated equilibria," yet often overlooked by population geneticists who focus largely on stability, rapid evolutionary changes occur mainly in populations far from equilibrium—within small, peripheral, "founder" populations obeying Darwin's appropriated image of "Malthusian instability," wherein self-replicators can grow exponentially provided that resources and fuel are plentiful. The organization of those raw materials and the mobilization of that free energy themselves comprise evolutionary adaptations, yet it is life's non-equilibrium environmental conditions, like the diverging cosmic conditions before, that establish the driving force essential to macro-evolutionary dynamics.

Although no commonly accepted criterion exists for biological complexity, non-"junk" genome size is one possible, very general measure (Szathmary and Maynard Smith 1995): Eukaryotes have larger coding genomes than prokaryotes, higher plants and invertebrates have larger genomes than protists, and vertebrates larger genomes than invertebrates. Variety of morphology and flexibility of behavior might provide other useful metrics of biological complexity (Bonner 1988): Higher life forms display a richer spectrum of bodily structure and a wider diversity of actions. Number of cell types in an organism might also be taken as an approximate indicator of its complexity (Kauffman 1991): Compartmentalization and specialization clearly rise in proportion to the number of genes in an organism, from bacteria to humans. Physical sizes of organisms provide yet another estimate of complexity (McMahon and Bonner 1983): Cellular specialization is proportional to an organism's size, from microbes to whales, and is evident in the construction of all life forms. However, it is hard to quantify any of these attributes of life and thus move the dialectic beyond the fluff of mere words, although they all do abide (with some exceptions) by the common-sense notion of complexity rising along most lineages.

In terms both general (for there are qualitative exceptions; see Gould 1996) and similar to those used earlier (for we do want to push the envelope regarding quantification), the more advanced a species, the greater the complexity (or the information density) of that species. Increasing specialization surely allows an organism to process more information, or energy, and just as surely the adoption of multiple strategies likely promotes survival, factors that aid biological evolution's production of complex organisms. To note just one prominent example, the buildup of oxygen in Earth's relatively early atmosphere permitted mitochondrial (eukaryotic) life to extract through respiration nearly 20 times more energy from the sugars it uses as food (oxidation of glucose to $CO_2$ and $H_2O$ yields 36 ATP molecules) than did prokaryotic life via fermentation in the absence of oxygen (which yields a mere 2 ATP molecules)—clearly an important step toward enhanced energy flow, increased $\Phi_m$ values, and greater overall complexity. That the general trend of living systems increasing their $\Phi_m$ values over generations is imperfect will not deter us, for no useful investigation can proceed if it must first scotch every ambiguity or explain every exception. We shall graphically encapsulate some of these statements by employing once more our criterion for complexity of free energy rate density flowing through open thermodynamic systems. Here, the earlier estimates of $\Phi_m$ for living systems are refined and extended, thereby connecting our previously developed arguments of thermodynamic physics with those of contemporary biology.

Consider first plants whose value of $\Phi_m$ at the end of Chapter Three was shown to have a global average of 900 erg $s^{-1}$ $g^{-1}$. At the time, we drew attention (compare note 6 in Chapter Three) to the inefficiency of the photosynthetic process, which normally converts only ~0.1 percent of the total incoming solar energy falling onto a field (Lehninger 1975). This value of $\Phi_m$ is sufficient to organize cellulose, the main carbohydrate polymer of plant tissue and fiber, for a field of uncultivated plant life, hence reasonable for the great bulk (>90%) of Earth's untended flora. However, a more organized field of corn photosynthesizes nearly ten times more efficiently, and a highly cultivated field of sugarcane (one of the most efficient converters of sunshine into biomass) can be higher still; their values of $\Phi_m$ are in the range of 6000 and 10,000 erg $s^{-1} g^{-1}$, respectively. The most highly organized of these fields requires a higher free energy rate density, not only because the higher values of $\Phi_m$ are consistent with increased organization of the plants, but also because the enhanced organization of the field produced by modern agricultural techniques requires higher $\Phi_m$ values to maintain that organization. The latter factor represents an energy contribution of a cultural, technological nature, yet another phase of cosmic evolution toward which our analysis is proceeding.

An alternative way of treating the energy density of plants, and thus assessing their complexity, is to examine their energy yields when they or their incompletely decayed remnants ("fossils fuels") are burned (Halacy 1977). In this way, chemical energy stored within molecular bonds is released, fossil fuels being an example of energy effectively stored (like a battery) in organic matter trapped within rocks. The original energy input naturally comes from the Sun, is temporarily captured in the plants during photosynthesis, and subsequently outputs heat when a torch frees up the energy stored in the plant, thus completing the flow through the system. More than 90 percent of the energy needed to power modern civilization derives from the decomposition products—coal, petroleum, natural gas—generated millions of years ago by photosynthetic organisms. The positive accumulation of free energy in these "fossilized" deposits represents a decrease in entropy locally at the expense of the Sun and are merely minor fluctuations in a general trend toward a state of higher global entropy. Take a field of hay, for example, for which ~42 tons will annually result from 1 acre of grain crop; the metric equivalent is $4 \times 10^7$ g and when burned yields ~$2 \times 10^{10}$ erg $s^{-1}$, making $\Phi_m \simeq 500$ erg $s^{-1} g^{-1}$. This is consistent with the above estimate for uncultivated plant growth such as grass, hay or alfalfa. By contrast, wood, which is basically glucose, is a somewhat more organized plant product, in which case a typical acre of softwood trees would yield ~$3 \times 10^6$ g (3.5 tons) per year. If burned, a softwood like pine would release (as the glucose bonds break) ~$10^{10}$ erg $s^{-1}$, making $\Phi_m \simeq 3000$ erg $s^{-1} g^{-1}$. In turn, hardwood would yield about double that, which is why mahogany burns "hotter" and often longer in the fireplace. These values are in keeping with the fact that a piece of fine wood has a higher degree of organization than a handful of loose grass.

In turn, high-quality coal, which of course is mostly fossilized organic plant product, can have a $\Phi_m$ value nearly twice that of hardwood; coal's order is said to be more concentrated, which is why its free energy rate *density* is larger.

Consider now a few representative animals for which metabolic rates are known, recognizing that those rates can vary upwards under stress and exertion, their total energy budgets dependent largely on energetically expensive internal organs such as kidneys, hearts, brains, and livers. Laboratory studies of sustained metabolic rates for 50 vertebrate species (Hammond and Diamond 1997) show that reptiles, mammals (including rodents, marsupials, and humans), and birds have average $\Phi_m$ values of 9,000, 56,000, and 78,000 erg s$^{-1}$ g$^{-1}$, respectively. These and other measurements strongly suggest that metabolic rates of ectotherms (cold-blooded organisms including all insects, fishes, amphibians, and reptiles) are only a fraction of those of similarly massive endotherms (warm-blooded organisms such as birds and mammals), much as we might expect on evolutionary grounds.

What is at first unexpected is that small birds have the highest specific metabolic rates among animals, especially during periods of peak activity, such as when earnestly foraging for food for their nestlings. Said another way, hummingbirds, for example, ingest food daily (mostly in nectar) up to nearly half of their body mass, compared to humans, who each day consume food in quantities equal to only a few percent of our body mass. However, birds, much like humans running a marathon, are fueled partly by consumption of bodily energy reserves (anaerobic glycolysis), not by sustained, concurrent energy intake. Upon second thought, flying creatures' large $\Phi_m$ values are neither surprising nor at odds with the central ideas of this book; after all, birds operate normally in a three-dimensional aerial environment, unlike much of the rest of animalia at the two-dimensional ground level, thus their *functions,* if not their structures, might be legitimately considered, somewhat and sometimes, more complex than those of the rest of us. Given that the birds (and the bees) have, while evolving the means ages ago to gather food, survive, and reproduce, also solved advanced problems in spatial geometry, materials science, aeronautical engineering, molecular biochemistry, and social stratification, then perhaps they *should* have large values of $\Phi_m$; in fact, any organism requiring an astonishing 160 flaps of its wings *per second,* as do big-bodied bumblebees during powered flight, would understandably have a large appetite. To be sure, swimming, running, and flying (let alone periodic reproduction and even social behavior) all represent biological functions that raise metabolic rates among animals; more energy is needed to perform and coordinate mechanical movements. We might even compare avian physiology to galactic physiology. In-flight birds, categorized among the highest complexity levels of animals, resemble "active" quasars within their class of galaxies, in that each tops the charts of metabolic rates (or light-to-mass ratios) with values of $\Phi_m$ well higher than their basal, or resting (normal) rates.

This brief book makes no significant distinctions among $\Phi_m$ values for members of the animal kingdom, except to note that they are nearly all within a factor of ten of one another, nestled nicely between those for photosynthesizing plants on the one hand and central nervous systems on the other. Also note that such values, surely among vertebrates and including humans, are not simply proportional to body mass, M. Measured basal metabolic rates themselves (in units of power) have been long known to vary as $M^{3/4}$, at least among different species of mammals whose body masses span some six orders of magnitude from skittering mice to placid elephants (Kleiber 1961). This empirical ¾ power, in contrast to the ⅔ power expected if metabolic rate scaled strictly as body surface area and if all animals were spherical, is potentially understood when animals are considered as elastic machines having muscular systems and skeletal loads subject to gravity (McMahon and Bonner 1983). Deep into discussions of biological metabolism we once more encounter a widespread astronomical factor—the gravitational force so integral in our earlier discussion of the underlying agents that established the conditions for increasing complexity. Astrophysics and biochemistry are not uncoupled parts of the cosmic-evolutionary scenario, as allometric scaling makes clear.

In cosmic evolution, as the present work contends, it is the *specific* metabolic rates that are most telling, namely those normalized to mass and which therefore for many life forms vary weakly with mass, namely as $M^{-1/4}$. This quarter-power scaling law is pervasive in biology, probably the result of physical constraints on the circulatory system that distributes resources and removes wastes in bodies, whether it is the geometrical pattern of blood vessels branching through animals or the vascular network nourishing plants. Natural selection seems to have optimized fitness by maximizing surface areas that exchange nutrients while minimizing transport distances and times of those nutrients (West, et al. 1999). That the smallest animals have the highest specific metabolic rates is often taken as an explanation of their frequent eating habits, somewhat high levels of activity, and relatively short lifespans; they live fast and die young. By contrast, the largest animals have the lowest specific rates owing to their more specialized cells, each of which has only limited tasks to perform and energy needed, thus granting greater efficiency and a longer life. Of course, those species whose individuals enjoy longevity are also likely to experience more extreme environmental stress and therefore be exposed during their lifetimes to enhanced opportunities for adjustment and adaptation, and thus for evolution toward greater complexity (as well as toward extinction should the stresses be too large). The result with the passage of time, as a general statement for bodies of similar size, is a feedback process whereby those successful systems able to assimilate greater energy flow live longer, evolve faster, and typically complexify, which, often in turn, leads to higher metabolic rates, and so on.

A mere extrapolation of this inverse dependence on mass does not explain why some respiring bacteria, rulers of the microbial world, have $\Phi_m$ values often reaching millions of times that of the Sun, or an astonishing $10^7$ erg s$^{-1}$ g$^{-1}$. Some microbes, such as the common soil species, *Azotobacter,* are the most voracious heterotrophs in Nature, producing 7 kg of ATP for each gram of its dry mass; a single microbe of the common bacterium, *E. Coli,* with a mass of $\sim 10^{-12}$ g and a doubling time of 22 minutes during exponential growth, would, if unchecked, produce a progeny of $\sim 10^{28}$ g, or roughly the mass of the entire Earth, within a single day! Perhaps this is why some species of prokaryotic bacteria are the most abundant—some would say, the most successful, even dominant (Margulis and Sagan 1986)—organisms in the biosphere. Aside from complexity measures, microbes might well be the "fittest" for their environment, for they invest a great deal of their stunning metabolism into reproduction. Yet the eubacteria (including life's most common prokaryotic bacterial groups) and their cousin archaebacteria (the most recently discovered, separate domain of life) are unicellular, unnucleated organisms only a few microns across, and that puts them in a class by themselves; creatures much smaller than the smallest mammals are not subject to gravitational considerations as noted above, since (surface tensional) forces other than those due to gravity and inertia become important for microscopic beings. Furthermore, being unicellular, bacterial cells are not specialized; each one must be general enough to accomplish the job needed to survive and replicate in a world of high surface-to-volume ratio. What's more, tiny cells are subject to life-or-death disordering effects that can harm them more readily, requiring more frequent repair; small structures like one-celled bacteria have more skin to maintain per pound than larger structures like huge elephants, for the same reasons that single-family homes have higher maintenance costs per pound than high-rise skyscrapers. All of which demands that the microbes have high energy flows per unit mass when respiring.

Microbes, however, like all else in this book, deserve to be placed into a larger perspective, for not all of them respire continuously; in fact few of them do. The aforementioned *Azotobacter* bacteria are indeed exceptional, having extraordinarily high respiratory rates far greater than for other aerobic bacteria. By contrast, more than three-quarters of all soil bacteria are virtually dormant, and thus have negligible $\Phi_m$ values, while eking out a living in nutrient-poor environments. Most of them enjoy a physiological ability to switch their metabolisms on and off, which is probably a survival-related trait. Accordingly, the metabolic scope, or range of $\Phi_m$ values above the basal rate, for microorganisms spans some ten orders of magnitude, a vast range nonetheless narrowed by environmental restraint; that such constraint is indeed operative among microbial life is exemplified by the presence of similar, yet different, microbes at Earth's two poles, almost surely the result of convergent evolution along quite independent paths (Morris 1998). Exhaustion of available resources as well as the accumulation of toxic products of metabolism are among the principal reasons that *E. coli* bacteria do not gobble up the whole Earth, however ridiculous that proposition may sound; their nutrients are severely limited. Although much of the microbial world remains the least explored part of biology, we hypothesize as before that, when the peak metabolic rates operating for short periods are time-weighted by the nearly negligible rates during much longer dormant periods, their average values of $\Phi_m$ range from hundreds to thousands of erg $s^{-1}$ $g^{-1}$, as expected for a system of such intermediate complexity. (An analogy might be made to a much larger life form, namely the world's biggest lizard. The Komodo Dragon of the Indonesian archipelago can consume up to 80 percent of its body weight at one meal, such as a 30-kg boar in 15 minutes, yet not need another meal for a month. Its time-averaged metabolic rate is a lot less than its instantaneous rate while eating.)

Among the eukarya (life's third domain that includes all plants and animals), the cold-blooded ectotherms have $\Phi_m$ values between 2000 and 10,000 erg $s^{-1}$ $g^{-1}$, whereas the warm-blooded endotherms have not only a similarly wide range of values but also higher absolute values, namely, $10^4$–$10^5$ erg s$^{-1}$ g$^{-1}$. The former are clearly among the earliest of biological evolution's animal creations, the latter widely considered more advanced, indeed among the most complex, of Nature's many varied life forms; with their portable microenvironments (shelter, fire, clothing, and so on), the endotherms have enjoyed a strong competitive edge, enabling them to adaptively radiate to even the most inhospitable parts of the biosphere. The just-quoted order-of-magnitude difference in specific metabolic rates among birds, mammals and comparably sized reptiles can legitimately be cast in terms of relative complexity, since the need for the endotherms to homeostatically control body temperature (both heating and cooling) is surely a more complicated task that the ectotherms simply cannot manage—and it is the extra energy that allows for this added feature, or selective advantage, enjoyed by birds and mammals over the past few hundred million years. Even so, the ectotherms are much more abundant, both as species and as individuals, meaning that they, too, are quite successful in their own more limited realms. As a qualifier, we ought not to belittle non-evolutionary effects that also contribute to the range in $\Phi_m$ values, for stressful environments do thermodynamically drive some species to extremes. One such extreme is represented by aquatic mammals, whose specific metabolic rates are necessarily two to three times higher than rates of similarly sized land mammals (since, again as for birds, they operate in a three-dimensional world, where icy water conducts heat 20 times faster than air). Another extreme is found in desert mammals, whose anomalously low specific metabolic rates reflect food shortages, though they can rehydrate rapidly by drinking the equivalent of a third of their body weight in 15 minutes. In any case, we do note a clear and general trend among macroscopic life forms between evolution-associated complexity and free energy rate density.

As for humans, our metabolic rates increase substantially by performing occupational tasks or recreational activities—again, that's function, not structure. At the end of Chapter Three, we derived a value of $\Phi_m \simeq 20,000$ erg s$^{-1}$ g$^{-1}$ for a rather sedentary, 70-kg adult, yet that was an operational rate—very much an average value while performing some minimal work (such as merely standing). Our basal rate, when lying in bed motionless several hours after eating, is some 60 percent less, or $\sim 11,300$ erg s$^{-1}$ g$^{-1}$. Various functions can increase that basal rate, and once again, it scales with the degree of complexity of the task or activity. For example, fishing leisurely, cutting trees, sewing by hand, and riding a bicycle require about 21,000, 58,000, 79,000, and

102,000 erg s$^{-1}$ g$^{-1}$, respectively (Smil 1999). Clearly, weaving an intricate pattern or balancing a moving bicycle are more complex functions, and therefore more energetically demanding tasks, than waiting patiently for the fish to bite. Thus we see how, in the biological realm, the value-added quality of functionality noted earlier does indeed count, in fact counts quantitatively.[3] To be sure, complex tasks performed by humans on a daily basis are typified by values of $\Phi_m$ that are a good deal larger than those of even the metabolically impressive birds, in part because birds cannot thread needles or ride bicycles!

Again, this short précis on cosmic evolution offers neither the proper time nor the appropriate space to sort through the huge and scattered database of thousands of energy costs tied to specific functional tasks, as measured in $O_2$-consumption experiments (or inferred from examining respiration chambers) by physiologists over the past century. Age and sex, leisure and labor, diet, health, stature, country, and education, among many other diverse factors affecting human life's highly varied energy budgets, result in surprisingly large individual differences in metabolic rates and subsequent free energy rate densities. Nor do we attempt to distinguish quantitatively between structural order and functional order. These are the "tasks," indeed ones requiring a great deal of research "energy," for a more comprehensive, "complex" study. Here we aim to stay on track to envision the bigger picture, unencumbered by the devilish details.

Figure 31 is the parallel for life forms of Figure 30, which graphs stars and planets at various evolutionary stages of matter. In the main, the trend is much the same for many of the major evolutionary stages of life: eukaryotic cells are more complex than procaryotic ones, plants more complex than protists, animals more complex than plants, mammals more complex than reptiles, and so on. Whether stars, planets, or life, the salient point seems much the same: The basic differences, both within and among these categories, are of degree, not of kind. We have discerned a common basis on which to compare all material structures, from the early Universe to the present—again, from big bang to humankind inclusively.

Let us be clear that the thermal gradients needed for an energy flow in Earth's biosphere could not be maintained without the Sun's conversion of gravitational and nuclear energies into radiation that emanates outward into unsaturable space. Were outer space ever to become saturated with radiation, all temperature gradients would necessarily vanish, and life among many other ordered structures would cease to

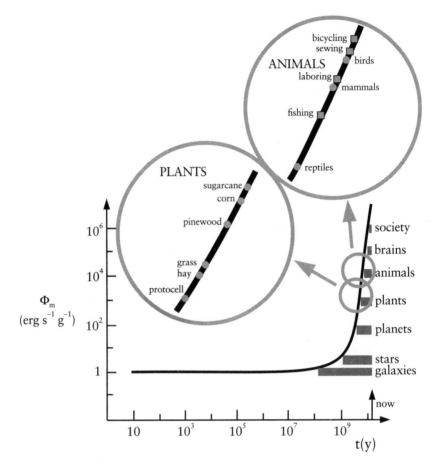

Figure 31. The earlier graph (Fig. 28) of rising complexity, expressed in terms of $\Phi_m$, the free energy rate density (reproduced at bottom), is here examined in more detail to highlight some of the ordered stages for a variety of animate objects discussed in the text—in this case, plant and animal life forms (square data points refer to humans only) in the biological-evolutionary phase of cosmic evolution.

exist; this is essentially a version of Olber's paradox, a nineteenth-century puzzle inquiring, in view of the myriad stars in the heavens, why the nighttime sky is not brightly aglow. That space is not now saturated (or the nighttime sky fully illuminated) can be attributed to the expansion of the Universe, thus bolstering the suggestion that the dynamic evolution of the cosmos is an essential condition for the order and maintenance of all organized things, including not only Earth, but also life itself. All the more reason to welcome life within our cosmic-evolutionary cosmology, for the observer in the small and the Universe in the large are not disconnected.

## Evolution: A Cultural Perspective

Energy has been central to our thesis throughout, but not just energy. Energy *flow* needs to be stressed, for without flow nothing much happens. Cosmic evolution is an active process that relies heavily on change; to be intentionally redundant, it is an *evolutionary* process. And there is probably no better documented example of the multifaceted dynamical flow of a system's energy than in our own biosphere, where the incoming Sun's rays (and its entropic tendencies) mix materials on Earth more thoroughly than might be imagined. Geologists have estimated that if a single thimbleful of water were poured into a river, after only a few years the circulation on a global scale would be so complete that a similar thimbleful of water taken from anywhere on Earth would contain some molecules from the original thimbleful. Others have expressed much the same idea regarding atmospheric circulation by estimating that every breath of air we inhale contains molecules that were once breathed by Galileo, Aristotle, and the dinosaurs.

That's mixing; that's entropy—the propensity to equalize, to achieve the lowest energy state, to spread things around evenly, in short to reestablish equilibrium at any and every opportunity. Water, air, and other climaspheric cycles can be very effective in redistributing energy and resources, some of the cycles impressively self-sustained when powered by energy, all in accord with the second law. Yet life itself defies the normal entropy degradation of the world around us; ironically, both Earth's life and Earth's pollution derive from the same principles of thermodynamics. The act of living is a kind of intermediate, improbable enterprise precariously balanced between the energy source (mainly) in the hot Sun and the energy sink in the cool Earth, and ultimately in cold, empty space. Again, as for all open systems, it is not energy as such that makes life go, but the flow of energy through living systems.

With the emergence less than a billion years ago of multicellular organisms as a new type of structure—whether individual plants, animals, or microbes—a novel degree of complexity had evolved. Levels of organization higher than that would require interactions beyond individual life forms, and ample evidence has accumulated that symbiosis did indeed occur on our planet, and is perhaps continuing (Margulis

1970). Communities of living systems along with their immediate environments, or ecosystems, in turn coexisted and coevolved, once again aided and abetted by robust flows of energy and considerably larger values of $\Phi_m$. Diverse species within such ecosystems carved out their own niches and different components of the ecosystem became increasingly coupled, as competition and reproduction among populations played active roles in the balance of Nature (Lotka 1924). In some cases, keen energy optimization and efficient material cycles have become well established, as is the case today, for example, in the Amazon basin where as much as 75 percent of the water is recycled. The importance of the input and output of matter and energy in Earth's food networks has long been recognized as a vital step in the organization of any ecosystem (Odum 1983). The net effect of increasing biodiversity—yet another name for the complexity of life forms en masse—within mature ecosystems is obvious (Wilson 1992). As with chemical evolution that led to primitive life and biological evolution that fashioned some of the first living ecosystems (or even stellar evolution within ecosystems of mature stars, interstellar clouds, and supernova remnants), communities of life forms (such as plants, plankton, and fish in a lake) exploit energy in two ways: The available energy in the area helps to establish the enhanced complexity initially, and once matured, a similar flow of energy helps to maintain the ecosystem as a dynamic steady-state in a geological, climatological, and biological setting. Such a balanced ecosystem is a good example of emergent behavior that cannot be explained without treating the entire process of which it consists.

The past billion years of evolution have caused our biosphere to become populated with a multitude of ecosystems, each exquisitely adapted to different environmental conditions and each having the means to sustain ongoing "traffic patterns" of incoming resources, outgoing wastes, and distributed energy, all so intricate and so optimized that a human-made replica of any one such system still eludes us. Day in and day out, these mostly ancient ecological systems continue on quietly, efficiently, and cleanly, each self-controlled by the steady flow of energy through Earthly environs. With occasional stress to their surroundings, ecosystems manage to transform and regenerate, to respond to the checks and balances of Nature, to self-organize once more as did less complex systems before them. What the untrained eye sees as disorder and randomness in an overgrown woodlot, a dense rain forest, or a backyard stream are actually splendidly ordered and impressively complex ecosystems. More than that, these are the life-support systems essential to humankind's survival. We are indeed entering the Life Era and the resulting landscape might be as new to us, and the ecosystems surrounding us, as it was to the newly formed atoms that began transiting into the Matter Era more than 10 billion years ago.

On a larger and more complex scale than individual ecosystems is the global biosphere itself, arguably the integrated sum of all Earth-based ecosystems. Others suggest that it is more than that, namely, that the whole biosphere is greater than the sum of its parts, the essence of holism once again. Modern-day Gaians—disciples of the Gaia hypothesis (Lovelock 1979) and descendants of worshippers of the terrestrial Greek goddess—go even further, claiming that planet Earth is a single, vast superorganism, indeed is itself alive ("strong Gaia"), a view not inconsistent with our expansive definition of life. Reality or metaphor, the idea of Gaia contends that living creatures affect the composition of Earth's atmosphere (and not just the other way around), both environment and life acting in a coupled way so as to regulate the former for the benefit of the latter ("weak Gaia")—actually an old notion dating back well more than a century and one that need not be at odds with neo-Darwinism. The microorganisms, for sure, excrete metabolic products that modify their environmental conditions, a property that has clearly enhanced their own species' survival, which not coincidentally extends, and most impressively so, over billions of years. Accordingly, living things have seemingly prevented drastic climatic changes throughout much of Earth's history, as evolution has endowed organisms with improved ability to keep surface conditions favorable for themselves. For example, as our Sun ages, thus becoming more luminous and sending more heat toward Earth, life responds by modifying Earth's atmosphere and surface geology to keep the climate fairly constant—a kind of "geophysiological thermostat" that basically adjusts $CO_2$ levels down by preferentially getting wrapped up in calcium carbonate under high temperatures, thereby cooling the atmosphere; if those temperatures go too low, the reaction to form carbonate decreases and the amount of $CO_2$ and the atmospheric temperature both rise, the whole $CO_2$ cycle acting as positive feedback that keeps water liquid and Earth habitable. And if the biosphere can be considered at least as a self-regulating and perhaps even self-reproducing system, then might not the concept of Gaia extend to the stars and galaxies beyond? By thinking broadly enough, one could conceivably reason that stars, too, are reproductive; as noted earlier, from the demise of some stars come the origins of other stars—a cyclical, generative (and not just developmental) change from galactic clouds to nascent stars to exploded supernovae, and so on through many generations, the whole process repeating endlessly in a kind of cosmic reincarnation. If this further blurs the distinction between animate and inanimate systems, then we are achieving another of our objectives. Revisionist views of Nature are also part of cultural evolution.

With the onset of human-induced problems of a global nature—atmospheric pollution, ozone depletion, overpopulation, scarce food and natural resources, and species extinction, to name several—the biosphere not merely remains the environment *for* society but has also now become an integral part *of* society (Csányi 1989). By any evolutionary standard, such technologically driven changes are extraordinarily rapid, so much so that the biosphere no longer seems able to respond well to the assaults of humankind. Even if the Gaia concept is correct and the atmosphere is able to resist perturbations, the time scale for such repair is far slower than human-induced change. The result would appear to be impending environmental crisis. The hard realities of occupational complexity in an industrialized society are upon us. Civilization's increased influence on our immediate environment has been so sudden and great in recent years that the ecological conditions needed for normal operations of society do not seem to be provided by the naturally changing biosphere. Humankind is moving toward a time, possibly as soon as within a generation or two, when we will no longer be able to expect Nature spontaneously to provide for us the environmental conditions needed for survival; rather, society itself will have to generate artificially the very conditions of our own ecological existence. From the two, society and the biosphere, will likely emerge a socially controlled bioculture. Here the components become ideas, artifacts, technology, and humans, among all other living organisms on Earth—the epitome of complexity writ large in Nature.

Social and cultural evolution, at the opposite end of the evolutionary spectrum from galactic and stellar evolution, updates our trek along the arrow of time, bringing us to us, *Homo sapiens,* harborers of the most complex clumps of natural matter known anywhere. Not that we seek to imply or infer anthropocentrism, a sentiment already roundly dashed in the Prologue and at the end of Chapter Three, indeed which deserves repeating here. No evidence whatever implies that humankind is the end-point of cosmic evolution, nor are we likely the only sentient beings in the Universe. The latter is especially true in recent years with the observational verification of numerous planets orbiting stars well beyond our own Solar System, as well as the widespread recognition that subterranean vents can harbor sufficient energy to sustain life even in the absence of a parent star. Yet, while completing our triple-distilled synthesis of radiation, matter, and life in this doubtlessly incomplete inventory, our technological society is currently and undeniably positioned at the most complex part of the cosmic-evolutionary scenario, at least as far as we can now decipher Nature's countless, change-filled events of the past. We have reached the here and now of our story.

By cultural evolution, we denote the changes in the ways, means, actions, and ideas of societies, including the transmission of same (which some call "memes"; Dawkins 1976) from one generation to another. Human culture, in particular, is the sum total of human minds, often acting cooperatively over the ages. Foremost among the more utilitarian changes that helped make us cultured, technological beings were the construction and refinement of useful tools, the development and teaching of meaningful communication, the invention and practice of profitable agriculture, and not least the discovery and harnessing of controlled energy. All of these are essentially manufacturing advances that expend energy to do work, decrease entropy at the locality of the cultural change, and more than compensate by increasing the entropy elsewhere in the world. Thermodynamic lingo may be unfamiliar to the practicing anthropologist or cultural historian, but the fundamental processes governing the changes in sentient life are much the same, albeit a good deal more complex, as for stars, planets, and primitive life.

Furthermore, these cultural traits were advanced and refined over scores of generations. The bulk of this new-found knowledge was transmitted to succeeding offspring, not by any direct genetic inheritance, but by use and disuse of information available indirectly in the surrounding environment. Without enumeration, imagine the increase in complexity and sophistication gained gradually or episodically during no more than the most recent million years: implements, language, foodstuffs, and power, among many other cultural factors acquired on the road to humanity. Of all these, language was probably paramount, ensuring that knowledge and experience stored in the brain as memory could be accumulated by one generation and transferred to the next— which is why, for some neurobiologists, language and intelligence are virtually synonymous.

A mostly Lamarckian[4] process whereby evolution of a transformational nature proceeds via the passage of acquired characters, cultural evolution, like stellar evolution before it, involves no DNA chemistry and perhaps less selectivity than biological evolution. Culture enables animals to transmit survival kits to their offspring by nongenetic routes; the information gets passed on behaviorally, from brain to brain, from generation to generation, the upshot being that cultural evolution acts much faster than biological evolution. Genetic selection itself operates little, if at all, in these two evolutionary realms that sandwich the more familiar neo-Darwinism, in which selective pressures clearly dominate. Even so, a kind of selection was at work culturally; the ability to start a fire, for example, would have been a major selective asset for those hominids who possessed it, an asset transmitted not by genes but by communication. It is this richness of factors and mechanisms in recent times that accelerates and complexifies change, which, in turn, confers a beauty and fascination to cosmic evolution.

As different as they are, biological and cultural evolution are not unrelated, as might be expected for two adjacent phases of cosmic evolution. Somewhat surprisingly, though, the two phases enjoy a subtle reciprocal interplay. Discoveries and inventions may well have been made by talented individuals having the "right" combination of genes, but once made, an invention such as lighting a fire or sharpening a tool would have, in turn, granted a selective (i.e., reproductive) advantage to those better endowed genetically to handle the skill. The two kinds of evolution thus partially complement one another, although in the recent history of humankind, Lamarckian (cultural) evolution has clearly dominated Darwinian (biological) evolution. Cultural acquisitions spread much faster than genetic modifications; our gene pool differs little from that of a Cro-Magnon human of 15,000 years ago, yet our cultural heritage is a good deal more robust in the knowledge, arts, traditions, beliefs, and technologies acquired and transmitted during the past 600 or so generations.

That cultural and social changes represent the most complex phenomena in the known Universe is undeniable. Human behavior, now engulfed by heavy energy use and rapidly changing environments, is what makes social studies so difficult. Unlike in much of the physical and biological sciences, controlled experiments in cultural evolution—humans interacting (social psychology), cities functioning (urban planning), or nations warring (political economics)—are virtually impossible to achieve. Just observing social behavior, let alone experimenting with it, is vastly more difficult to accomplish than manipulating molecules in a chemistry laboratory or sending space probes to the planets;

likewise, the number and diversity of factors influencing the outcome of a human interaction is far greater than those affecting the birth or death of a star. Although a physicist or chemist might never have a concern for an individual atom or molecule, sociologists often treat the human behavior of a single individual as paramount—and that, ironically, is what makes their task all the more difficult, and complex, for not even statistical reasoning can help much.

A few examples of Lamarckian-style cultural change toward greater complexity will suffice. Consider first the just-mentioned preeminent exemplar of culture: language. Language is transmitted largely through the media of teaching and schooling, passing on knowledge from adult to offspring, not perfectly but adequately (and including imitatively) over the years. Not only do we transfer information to our children in this way, but the body of available knowledge itself also grows with the acquisition of new stories, facts, and techniques. And because that knowledge accumulates faster than it is forgotten (especially with the onset of recorded history), the sum total of culture passed along builds, indeed grows richer and more complex. That's why it now takes a third of a human lifetime to train for a doctorate in science, despite deep specialization. Human knowledge today far exceeds that of any one individual. Hardly any of our cherished educational facts, models, and methods are transmitted via the genetic biochemistry of the genes; those genes do grant a hard-wired ability to learn from other human beings, but learning itself is a surprisingly long, hard struggle, often overcome by application of, yes, energy. The story of cosmic evolution is itself a cultural myth hereby disseminated in the form of this book—a myth, because it is admittedly a simplification of an exceedingly complex approximation of reality.

Industrial development is another cultural practice that decreases entropy locally in the form of artificially manufactured products, yet only with the sweat and toil of spent energy, which inevitably increases entropy in the larger environment of raw materials used to make the goods. Modern automobiles, for instance, are better equipped, sounder mechanically, and basically safer than their decades-old precursors, not because of any internal tendency to improve, but because manufacturers have constantly experimented with new features, keeping those that worked well while discarding the rest—a clear case of acquiring and accumulating successful characteristics from one generation of cars to the next. A kind of selection pressure functions by means of dealer competition and customer demand in the social environment—selection as human preference in the marketplace—the evolutionary mechanism being more Lamarckian than Darwinian. There is little doubt that Lamarckian tinkering can improve technology: Use and disuse engenders gradual change in automotive style, operation, and safety, all of which feed back to increase the pace of our lives and the thrust toward even more complexity. Would anyone deny that today's gadget-filled automobile (with computer-assisted fuel injection, electronic valve timing, and microprocessor-controlled turbochargers, not to mention all those widgets on the dashboard) is more complicated than Ford's "Model T" of nearly a century ago, or that more energy is expended per unit mass to drive them?

As a reality check on our thermodynamic analysis, in this case applied to cultural evolution, we calculate the free energy rate density for an average-sized modern automobile, whose typical properties are $\sim 1.5 \times 10^6$ g (or $\sim 1.6$ tons) of mass and $\sim 10^6$ kcal of gasoline consumption per day; the answer, $\Phi_m \simeq 10^6$ erg s$^{-1}$ g$^{-1}$ (assuming 6 hours of daily operation), is likely to range higher or lower by several factors, given the variations among the population of vehicle types and choice of fuel grades, yet this average value is very much in line with that expected for a cultural invention of considerable magnitude—indeed, for what some say is still the archetypal symbol of American industry. Put another way to further illustrate the evolutionary trends and using numbers provided by the U.S. Highway Traffic Safety Administration for the past two decades, the horsepower-to-weight ratio (in units of hp/100 lb) of American passenger cars has increased steadily from 3.7 in 1978 to 4.2 in 1988 to 5.1 in 1998; these values, when translated into the units of $\Phi_m$ used here, equal $5.9 \times 10^5$, $6.8 \times 10^5$, and $8.3 \times 10^5$ erg s$^{-1}$ g$^{-1}$, respectively. These numbers in and of themselves, and also when compared to the less efficient and sometimes heavier autos of yesteryear (whose values of $\Phi_m$ average less than half those above), confirm once more the general correlation of free energy rate density with complexity.

That correlation can be more closely examined by tracing the evolution of the internal combustion engine and gas turbine, notable examples of technological innovation during the power-hungry twentieth century. The four-stroke Otto engine of 1880 had an energy flow of $4 \times 10^4$ erg s$^{-1}$ g$^{-1}$, which was quickly surpassed by the multi-cylinder Daimler engine of 1900 ($\sim 2.5 \times 10^5$ erg s$^{-1}$ g$^{-1}$). More than a billion such engines have been installed to date in cars, trucks, aircraft, boats, lawnmowers, etc., thereby acting as a signature force in the world's economy during the past century. Today's mass-produced cars and trucks, as noted above, average several times the $\Phi_m$ value of the early Daimler engine, and some racing cars can reach an order of magnitude higher. Among aircraft, the Wright brothers' homemade engine ($\sim 10^6$ erg s$^{-1}$ g$^{-1}$) was superseded by the Liberty engines of World War I ($\sim 7.5 \times 10^6$ erg s$^{-1}$ g$^{-1}$) and by the Whittle–von Ohain gas turbines of World War II ($\sim 10^7$ erg s$^{-1}$ g$^{-1}$). Arguably today's premier example of mass transportation, the Boeing 747 jumbo jet, with 110 Megawatts powering an 80-ton craft to barely subsonic velocity, displays an average $\Phi_m = 2 \times 10^7$ erg s$^{-1}$ g$^{-1}$. As for modern jet fighters, the Phantom F-4 of the 1960s and the Nighthawk F-117 of the 1990s have impressively larger values of $\Phi_m$, $3 \times 10^8$ and $8 \times 10^8$ erg s$^{-1}$ g$^{-1}$, respectively, testifying to the increased complexity of these supersonic and sophisticated machines (Smil 1999).

Our final example of cultural evolution, among many others that could be chosen, deals with the icon of our modern information society: the computer chip and its advances in memory capacity and data processing speed—miniature silicon-based widgets that, by our criterion of free energy rate density, are stunningly complex. In particular, the number of transistors—small semiconductors used as electrical amplifiers and logic gates—that fit within a single computer microprocessor has increased greatly over the past few decades, in fact doubling geometrically about every 18 months ("Moore's law"); today's powerful Pentium-II chip holds more than 1000 times as many transistors (7.5 million) than the famous Intel-8080 chip (6000 transistors) in the pioneering days of personal computers a quarter-century ago. Stated alternatively, the rate of energy density flowing through each transistor has increased in recent times, the result of culturally acquired knowledge accumulated from one generation of computers (and their engineers) to the next, a decidedly Lamarckian process. For these calculations, we took the power input to the 8080 and Pentium chips to be 1 W and 10 W, respectively, and the mass of the 4-mm $\times$ 4-mm $\times$ 2-$\mu$m chips to be $\sim 1$ mg, making $\Phi_m \simeq 10^{10}$ and $10^{11}$ erg s$^{-1}$ g$^{-1}$.

Another way of expressing such high densities is to note that in 1999 a typical 0.2-m (8-inch) silicon wafer holding several hundred chips—the soul of computer manufacturing—contains approximately 10 billion components, which is more than all the people on planet Earth. Our so-called post-industrial society may well have already constructed more transistors than any other product in human history, including bricks! So, although the power consumed per transistor has decreased with the evolution of each new generation of chips, the energy rate *density* has continued to rise because of miniaturization and its consequent complexification. Much of the high $\Phi_m$ values for modern computer chips derives from the size of the micron-scale transistors, many of them having individualized and ultra-miniature fans to dissipate their growing heat. A modern desktop computer can do more things, and quicker too, than could an old room-sized mainframe because the former is more complex; the central processing units of today's personal computers have immense throughputs of both information handling and free energy rate densities, a not altogether surprising result since these two quantities are clearly related. Furthermore, the chips' $\Phi_m$ values exceed those of human brains because computers number-crunch a lot faster than do our neurological systems. That doesn't make today's microelectronic machines more sentient than humans, but it does make them more complex, given their extraordinary functional data-processing speeds. On the drawing boards now are advanced technologies for increasing the number of transistors on a single silicon chip to the order of a billion, probably along with a continued rise in $\Phi_m$, all apparently attainable within the early decades of the twenty-first century. Our information age will then likely continue rocketing upward, either successfully to achieve new heights of complex, artificial things or self-destructively to ends unknown—either path an entropy-producer, the result of human-driven and Lamarckian-style cultural evolution.

Our treatment of culture, peculiar though it is from a quantitative perspective, need be addressed no differently than for any other aspect of cosmic evolution. Free energy rate densities can be calculated by thermodynamically analyzing society's use of energy among our relatively recent ancestors (Cook 1971). That the increasing expenditure of energy has been an important factor in the emergence and development of humankind is a given; cultural evolution occurs when societies alter their organizational posture in response to changes in the flow of energy through them, for it is increased organization that helps dynamically stabilize a far-from-equilibrium society. The most primitive societies of hominids had available for work only the physical energy of their individual work ethic; for a 50-kg post-australopithicene forager of a couple of million years ago, that energy would have been approximately 2000 kcal of food eaten per day, granting it a $\Phi_m$ value of about 19,000 erg s$^{-1}$ g$^{-1}$. With the controlled use of fire some hundred thousand years ago, the exploitation of energy would have roughly doubled, making $\Phi_m \simeq 40,000$ erg s$^{-1}$ g$^{-1}$ for a slightly heavier *Homo habilis* creature. The poverty of energy still somewhat limited cultural development, but with the onset of agricultural society and the use of domesticated animals some 10,000 years ago, the equivalent energy available to individual *H. sapiens* (now considered for these calculations to be the model 70-kg male) increased to about 12,000 kcal day$^{-1}$, making $\Phi_m \simeq 83,000$ erg s$^{-1}$ g$^{-1}$; these, in turn, would have easily doubled with the invention of advanced farming techniques and the emergence of metal and pottery manufacturing a few millennia ago. Much more recently, the start of the Industrial Revolution and especially the burning of coal afforded each member of a young, mechanistic society (especially in Britain, Germany, and the United States) a great deal more energy for use in daily, societal activities; we compute $\Phi_m \simeq 480,000$ erg s$^{-1}$ g$^{-1}$.

These increases in $\Phi_m$ values have been evolutionary, competitive processes in which selection has once again played a role. New technologies drove older ones to extinction, thereby benefiting humankind over the ages (Brown 1976). Throughout the past few centuries, for example, the "customer" chose shorter travel times, lower transportation costs, and heavier shipping loads; steam-powered iron ships replaced the wind-powered modern clipper ships. Likewise, "horsepower" provided literally by horse and mule was first marginalized and then eliminated by steam and eventually gas engines as work animals on most farms the world over; people elected to concentrate their energy for greater efficiency. Typewriters, ice boxes, and slide rules, among many other innovative advances in their own time, were selected out of existence by the pressures of customer demand and commercial profit, often replaced initially by luxuries that eventually became necessities, such as the word processor, refrigerator, and pocket calculator. But all of this progress, which did subjectively better the quality of life (as measured by health, education, and welfare), came at the expense of greatly increased energy consumption. Rising energy expenditure per capita has been a hallmark of cultural evolution, an idea dating back at least decades (White 1959), though without any attempt to address causality, as well expressed by the American anthropologist Richard Adams (1975): "To ask whether this increase in energy in the system is 'caused' by something or another is not helpful if one seeks a 'prime mover'; the fact that it conforms to the widely observed principle of natural selection and the Second Law of Thermodynamics will have to stand as an 'explanation' for the present. The concentration of energy within the life systems follows the Second Law, and such a concentration will continue insofar as conditions of life may be met and the continuing supply of energy from the sun is available." The present work seeks to specify, if broadly, that causative agent, or prime mover, in the guise of cosmic expansion which, in turn, advances the arrow of time.

The rise in free energy rate density has been steady into modern, technological times, as today's world has indeed become a humming, beeping, well-lit place. With centralized power generation and distribution plants increasing the effective daily usage of energy, the per citizen expenditure in industrialized countries averaged 230,000 kcal by 1970, or $\Phi_m \simeq 1.6 \times 10^6$ erg s$^{-1}$ g$^{-1}$. (Notice how the values for $\Phi_m$ for automobiles and aircraft, calculated a few pages prior, fit nicely into this spectrum of complexity.) Now, at the start of the twenty-first century, with 35 percent of the world's total power (~18 trillion watts) consumed by only 5 percent of the world's population living in the United States, our country averages $\Phi_m \simeq 3.2 \times 10^6$ erg s$^{-1}$ g$^{-1}$ (which amounts, by the way, to ~25 kW for each U.S. citizen, compared to ~3 kW per person globally). Thus, modern high-tech conveniences, from the aforementioned automobiles, aircraft, and computers to a wide variety of energy supplements enhancing our information-based society (including wired homes, networked businesses, and consumer electronics of all sorts), have empowered today's individuals well beyond their 2800 kcal of food consumed daily. Figure 32 plots many of these culturally oriented values of $\Phi_m$, thus conforming with our previous survey of open systems—of which today's technologically sophisticated society is just one more example.

Increasingly, the makeup and functioning of whole cities, states, and nations are beginning to be couched in terms of non-equilibrium thermodynamics, for these too are dissipative structures (Dyke 1988; Jervis 1997). Cities are dynamic steady-states, like any open system. They acquire and consume resources, produce and discard wastes, all the while processing energy for all manner of services: transportation, communications, construction, health, comfort, and entertainment, among a whole host of maintenance tasks. Modern cities are as much a product of an evolutionary process as any galaxy or organism, and many are still developing, seeking to establish dynamically stable communities within our planet's larger, vibrant ecological system. Their populations are dense, their structures and functions highly complex; cities are voracious users of energy, their values of $\Phi_m$ higher than any in the previous paragraph (all normalized to the mass of the system for valid comparison). In a way, it is too early to tell if cities will survive; they are

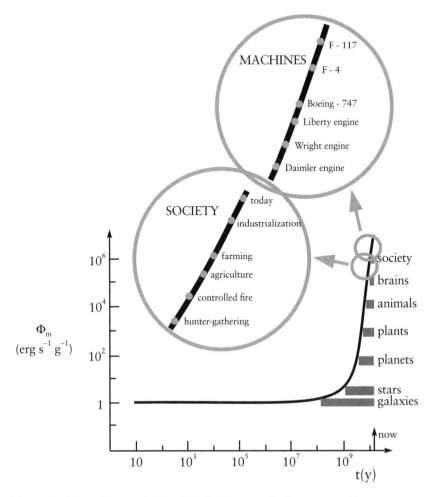

Figure 32. The earlier graph (Fig. 28) of rising complexity, expressed in terms of $\Phi_m$, the free energy rate density (reproduced at bottom), is here examined in more detail to highlight some of the ordered stages for a variety of human-related advances discussed in the text—in this case, consumption of energy by human ancestors and artificial gadgets in the cultural-evolutionary phase of cosmic evolution.

among the youngest advances in cultural evolution, prone to physical and social constraints that might well fundamentally change, or even eliminate (via selection), those very same cities. Urban planners could do worse than an integrated, evolutionary analysis of the social and environmental challenges now confronting human settlements.

Economies, too, are products of evolution; they are modes of organizing ecological space for increased flux, enhanced dissipation, and greater productivity. Although orthodox economic theory—even that which embraces thermodynamic thinking (Georgescu-Roegen 1971)—treats the action of goods exchange as if it were reversible and equilibrated, much could be gained if economies were modeled as open systems enjoying a far-from-equilibrium status (Day 1987; Ayres 1994). The emerging interdisciplinary subject of ecological, or evolutionary, economics highlights the celebrated concept of energy flow (including material resources), just as other bastions of specialization have rallied around energy flow as central to the interdisciplines of astrophysics and biochemistry. Understandably, social scholars concerned about natural science treading on their turf will at first reel at the notion of non-equilibrium conditions, market gradients, and institutional flux, all implying economic life (and politics) on the ragged edge of chaos. Yet if we have learned anything from the foregoing analysis, it is that all organized entities exist uneasily "on the edge," from unstable giant stars to struggling life forms to endangered ecosystems. It is, again, their dynamic steady-states that act as sources of innovativeness, creativity, and the very way that systems take advantage of chances to advance steadfastly along the scale of complexity. That mixture, once more, of randomness and determinism is also why realistic economies will never be predictable in detail, but will remain process-dependent, dynamic, and always evolving. By contrast, with regard to nation-states and the financiers who seek to control them, economic equilibrium would signify a meltdown, indeed a "heat death," of modern society—the unequivocal collapse of global markets.

Throughout the past million years or so, biological and cultural evolution have been inextricably interwoven. Their interrelationship is natural, their overlap consistent with other adjacent phases of cosmic evolution that also interact, for the development of culture bears heavily on one of those key factors affecting all of evolution—the environment. Cultural inventiveness enabled our immediate ancestors to evade some environmental limitations: Hunting and cooking allowed them to adopt a diet quite different from that of the australopithecines, clothes and housing permitted them to colonize both drier and colder regions of planet Earth, and tools allowed them to manipulate their localities, however primitively. As for biological organisms before them, specialization permits social organizations to process more information and utilize more energy.

Likewise, though more dramatically, present cultural innovations enable us—twenty-first-century *H. sapiens*—not merely to circumvent the environment but also to challenge it directly. Technology allows us to fly high in the atmosphere, to explore deep within the oceans, even to journey far from our home planet. Change indeed now quickens and with it the pace of life. Culture and its most common currency—energy, front and center—apparently act as catalysts, speeding the course of change toward an uncertain future. Humankind is now largely in charge of life on Earth; controlling most events, making key decisions, doing the selecting. We have become the agents of change, the human drivers of cultural evolution.

If there is any one factor that has most characterized the evolution of culture, it is almost surely an increasing capacity to extract energy from Nature—but not merely to capture energy, rather to store it, to transfer it, in short to process energy. Over the course of the past 10,000 years or so, humans have steadily mastered wheels, agriculture, metallurgy, machines, electricity, and nuclear power. Soon, solar power will emerge in its turn; all intelligent civilizations, anywhere in the Universe, likely learn to exploit the energy of their parent star. Each of these innovations has channeled greater amounts of energy into culture, in fact increased the rate of energy density flowing through many an open Earth system that serves us daily. To be sure, the ability to harness abundant energy sources is the hallmark of modern society. But it is also clearly the source of an inexorable rise in entropy within our larger environment—widespread pollution, waste heat, and social tumult, among other societal ills. Ironically, the use of energy and natural resources so vital to our technological civilization is also a root cause of many sociopolitical problems now facing humankind at the dawn of the new millennium.

## Summary: A Cosmic Perspective

Physical, biological, and cultural evolution span the spectrum of complexity, each comprising an essential part of the greater whole of cosmic evolution. Stars, planets, and life, as well as culture, society, and technology, all contribute magnificently to a coherent story of ourselves, our world, and our Universe. All these systems, among many other manifestations of order and organization, here and elsewhere, seem governed by common drives and attributes, as though a Platonic ideal may well be at work around us. At all times in the Universe, and at all places, the second law of thermodynamics is the ultimate arbiter of Nature's many varied transactions; it, and the ubiquitous process of energy flow directed by it, embody the underlying physical principle behind the development of all things.

Earth is in the balance. Our planet constitutes a precarious collection of animate and inanimate localized systems amid a complex web of global and cosmic energy flows. All these systems—whether entirely natural or engineered by humans—need to heed the second law as an unavoidable ground rule. Consciousness, too, including societal planning and technological advance likely to dominate our actions into the next centuries, must embrace an evolutionary, thermodynamic outlook, for only with an awareness and appreciation of the bigger picture can we survive long enough to experience the Life Era, thereby playing a significant role in our own cosmic-evolutionary worldview.

EPILOGUE

# A GRAND SYNTHESIS

*. . . all tangible phenomena, from the birth of stars to the workings of social institutions, are based on material processes that are ultimately reducible, however long and tortuous the sequences, to the laws of physics.*

—Edward O. Wilson, *Consilience: The Unity of Knowledge*

Many in human history have undoubtedly reasoned that they lived in fortunate, interesting times. Despite much physical suffering, political injustice, and social inequity among members of our species, the phrase "it's a wonderful life" may have been a common proverb for much of humankind ever since our ancestors became conscious of themselves. Broadly considered, as has been our aim throughout this book, life does seem to be a truly wondrous phenomenon even if, in our modern day and age, largely devoid of mystery. Not necessarily unique to Earth, the phenomenon of life is manifestly remarkable when examined in its larger, cosmic setting. At the start of the third millennium, those of us called to science (and not merely professional scientists) do now partake of exciting, productive, indeed wonderful times. Natural science has seldom been more robust, acquiring much new data, establishing heaps of novel facts, generating elegant theoretical insight, spearheading our technological civilization, and renewing our appreciation for the nature of the Universe as well as for the life within it.

We are now living in a golden age of astrophysics. Our great-grand-children will someday judge the last quarter of the twentieth century to have been a period of especially fertile adventure and discovery in the world of science. Not since Galileo and Newton have the physical sciences so revolutionized our perception of the Universe. A single generation—neither that of our parents nor that of our children, but *our* generation—has opened for exploration the entire electromagnetic spectrum, mining it for rich databases and wider understanding. Access to space has helped fuel our scientific fortune, as has the assembly of powerful computers, yet so has the sea-change toward inter-disciplinarity that grants context and perspective so vital to overall comprehension. The result of these advances in physical science is unprecedented intellectual grasp of our planet, our star, and the cosmos.

Likewise, historians of the future will surely conclude that we also now share in a golden age of biochemistry. Not since Darwin have the biological sciences accelerated so dramatically, especially in today's burgeoning field of molecular biology. The intense pace and penetrating insight from rapid breakthroughs in the biological sciences are equally impressive as those of the physical sciences. The unraveling of life's genetic code, stunning discoveries in modern medicine, invention of powerful techniques in biotechnology, and a renewed respect for biodiversity, all herald a healthy vigor within the burgeoning bioinformatics community. The result of these advances in biological science is extraordinarily detailed knowledge of life, culture, and our human selves.

What's more, the interplay and perhaps merger of the physical and biological sciences are yielding a unified worldview—an integrated *Weltanschauung*—and bolstering it with unsurpassed empirical information, amid continuing adherence to the Renaissance-inspired scientific method. That unification we have called "cosmic evolution," for it is the phenomenon of change that is paramount on a truly universal stage. In particular, we have examined the gradient-rich conditions conducive to the growth of complexity, explored myriad changes occurring on virtually all spatial and temporal scales, emphasized the importance of energy flowing through open, non-equilibrium systems, and identified many common denominators among, frankly, all organized things. We have been especially keen to decipher, while studying the widest possible range of structures, how order has emerged, indeed increased, within localized parts of the Universe, such that galaxies, stars, planets, and life forms are but islands of complexity within a vast sea of disorder that is the cosmic environment beyond.

The intent was to examine to what extent all ordered structures could be studied "on the same page," to break down the barriers separating the natural sciences and to probe any coherent similarities, guiding principles, or grand syntheses connecting a wide spectrum of systems otherwise seemingly unrelated—imposing systems like stars, life, and society. Our exposition has admittedly sought an a priori objective, a test of a preconceived idea, while adopting only mainstream science as a tool—good, accurate, quantitative science, though often liberally interpreted and broadly explained. In utilizing the most universal currency in modern science—energy—we have found a reasonable way to compare compact stars and budding plants, active galaxies and diverse cultures, including ourselves as sentient beings, thereby achieving a certain level of analytical detail and computational measure, all the while asserting only that this condensed précis might contribute not truth but merely a better approximation of reality.

Energy flowing through individual systems might seem like a trivial proposition for a pervasive organizing principle in Nature, given energy's manifold roles within the often disparate disciplinary sciences. Yet that is the point: When the specifics of energy *change* are examined (and normalized) among all organized systems in an interdisciplinary fashion, this concept appears much more powerful and comprehensive, with energy itself acting as a driving-force dynamic throughout all of Nature writ large. The values of free energy rate density ($\Phi_m$) derived herein span the utmost landscape, containing stars and galaxies, plants and animals, societies and machines, literally enabling us to address the widest known spectrum of organized entities with the same set of diagnostic tools. The objective has been, however grandiose it may sound, a unification of the accepted ideas and observable results of much of modern science, based on common principles and repeatable experiences. To give credit where due, Heraclitus (and perhaps others before him) basically had it correct when arguing long ago that the primary substance is fire and that everything changes—παντα ρει, an ancient Greek maxim, meaning "all flows." With cosmic evolution, modern science not only roundly welcomes that old ideal but also now embraces a newly invigorated Heraclitean ethos. To be sure, when considered in the context of non-equilibrium thermodynamics, and without encountering any compelling need to invent some new law that supersedes the second law of thermodynamics, the flow of energy does seem to be a reasonable candidate for an underlying phenomenon that integrally affects all things.

Not least, we have also been guided by notions of beauty and symmetry in science, by the search for simplicity and elegance, by an attempt to explain the widest range of phenomena with the fewest possible principles. The energy arguments in this book, even when bolstered with quantitative analysis, are as much architectural as they are technical. Like an architect who uses technical information to serve a larger purpose, herein we have adopted energetics to weave an expansive yet intricate tapestry depicting the origin and evolution of all material structures. The added cogency gained from the construction of such a grand and portentous worldview implies, in and of itself, a whole exceeding the sum of its many components. The resulting evolutionary epic, even when triple-distilled as in this brief work, rises above the collection of its copious parts, potentially granting meaning and rationality to an otherwise unworldly endeavor. Intelligent life is an animated conduit through which the Universe comes to know itself.

Biochemistry does have much in common with astrophysics; life apparently is an integral part of the cosmos—a cosmos simultaneously nurturing and indifferent, complexifying and disordering. Without clear evidence for any design or purpose anywhere in Nature, humankind does seem to have naturally achieved a level of sentience able to know of our past if not our future, to appreciate the myriad changes that have brought us life, to realize that we are children of the Universe at the dawn of the Life Era. A consistent *and natural* case can indeed be made for a cosmic-evolutionary worldview that goes beyond the mere words of poetry and the superficiality of metaphor—a worldview incorporating empirical measures of origins, ages, energetics, structures, and perhaps meaningful understanding.

Specialized research held sway in the science community for much of the twentieth century, providing theory, data, and occasional synthesis in an ongoing search for fundamental knowledge about ourselves and the Universe—in short, the focused scholarship needed to fatten the skeletal pattern of intellection that upholds the profound narrative of cosmic evolution. To be sure, that scholarship has led to many practical applications of science that have also been the hallmark of a productive and beneficial economy for much of the last half-century; for that profit-oriented reason alone, specialization in science and its associated technologies will continue unabated. Yet, like much of Nature, change is afoot even in the world of science, where the philosophy of approach is also fluxing, swinging like a pendulum toward a more balanced posture, away from strict reductionism and toward unifying holism. Interdisciplinarity is today more fashionable, and valuable, in science—crossing traditional boundaries, networking among fragmented disciplines, and generally welcoming a more synergistic approach to problem-solving.

Perhaps now is the time to widen the quest for understanding still further, to expand the intellectual effort beyond conventional science—to engage the larger, non-scientific communities of philosophers, theologians, and others who often resonate with the cosmic-evolutionary theme even if not in name, all in an ambitious attempt to construct a millennial worldview of who we are, whence we came, and how we fit into the cosmic scheme of things as wise, ethical human beings.

Humankind is entering an age of synthesis such as occurs only once in several generations, perhaps only once every few centuries. The years ahead will surely be exciting, productive, perhaps even deeply significant, largely because the scenario of cosmic evolution provides an opportunity to inquire systematically and synergistically into the nature of our existence—to mount a concerted effort to create a modern universe history *(Weltallgeschichte)* that people of all cultures can readily understand and adopt. As we begin the new millennium, such a coherent story of our very being—a powerful and true myth—can act as an effective intellectual vehicle to invite all citizens to become participants, not just spectators, in the building of a whole new legacy.

## Frequently Asked Questions

Numerous questions have inevitably arisen while discussing the subject of cosmic evolution with experienced faculty, beginning students, and lay persons alike. Here is a representative sampling of those queries most frequently asked, presented to help clarify our central arguments and to avoid lingering misconceptions.

### Is the Cosmic-Evolutionary Scenario Complete?

Our treatment of cosmic evolution set forth in this book is by no means complete or comprehensive, especially regarding the devilish details. Cosmic evolution is basically an historical science and, like most histories, it will remain a partial, if increasingly fertile, record of the past. We share the lesson familiar to all historians: Rarely is it possible to recreate (even in the mind's eye) the precise environments in which distant events occurred, yet it is often adequate to infer from scattered relics of the past approximately what did happen. Since our vision of cosmic evolution "goes all the way back" in deep time, it is all the more deficient, pending inclusion of future discoveries and novel insights into the expansive, 12-billion-year pattern it portrays. Like thermodynamics, which is so powerful because it is so general, thereby restricting itself to the broadest of statements, cosmic evolution also appeals more to breadth than depth, yet without apology, for its wide vista is its true forte. This brief account of cosmic evolution has sought mainly to identify common features—principally energy—bridging Nature's untold ordered structures, and then to track the changes in that energy over the course of all time. The result is a rich though incomplete narrative encompassing an impressive expanse of universal real estate, indeed a grand synthesis of an extensive inventory of radiation, matter, and life, each comprising an integral part in the search for unity of knowledge.

*Is Cosmic Evolution Mere Metaphor?*

A metaphor is the application of a word or phrase to a concept it does not literally denote, in order to suggest comparison with another concept. Our broadened view of evolution, as in cosmic evolution, is metaphorical only if all of science is a metaphor for general understanding. Given the powerful underlying phenomenon of change quite naturally everywhere, evolution itself should not be a disciplinary word exclusive to only one field of science, but rather an interdisciplinary word that helps connect often disparate fields of scientific scholarship. There is no reason whatever, given the definition of the term offered in the Glossary, that evolution cannot be used broadly to describe all the many varied changes observed throughout Nature. As such, neo-Darwinism, which has largely appropriated the term for itself, becomes but a special case (with powerful value-added features) within the much wider purview of cosmic evolution. At least we have resisted calling cosmic evolution a new paradigm; its central ideas hark back decades, if not centuries. The term "scenario" best denotes a working hypothesis, or maturing narrative, for cosmic evolution does summarize in a neat and testable intellectual package what is otherwise the messy and perplexing materiality of Nature. Is this story the truth? We wouldn't know, for who can define truth? As we progress in understanding ourselves and the Universe around us—and we do progress; see the relevant question below—we gain an increasingly better approximation of reality. And it is that critically appraised reality that enables us to compose the materialist worldview presented herein.

*How Does Energy Flow Drive Evolution?*

Energy flow can take the sting out of time, thereby taming its degrading influence. Where the environment is conducive and energy available, open systems capable of utilizing energy can sometimes build more complex structures able to process increased energies; the products often comprise an unresolvable mixture of efficiency, competition, and optimization. What's more, the resulting complexity is not necessarily just an accumulation of more of the same; rather, increasing amounts of it can sometimes lead to distinctly new (emergent) phenomena. The complex objects seen around us are the macroscopic manifestations of fundamental fluctuations—that's largely chance—arising on the microscopic level within unstable dynamical systems. A combination of nonequilibrium conditions and resultant system-environment gradients engender energy flows—that's largely determinism—not to achieve some predestined goal since evolution is not a movement toward any specific state, but to foster ordered forms consistent with the easiest way out of a difficult situation. In turn, via trial, error, and much feedback, evolution improves the management of energy flow with each substantial increase in organization, all in accord with a selection process that is quite natural, thereby rewarding and reinforcing those flow pathways able to draw greater rates of energy per unit mass.

## Are Complexity and Energy Rate Density the Right Terms?

We have been intentionally liberal with the term "complexity," using it as one of those common words threading our cosmic-evolutionary narrative, yet realizing that others may prefer to describe the state and degree of order by alternative names. In addition to information content and negentropy that were earlier used for qualitative appreciation (and then discarded as quantitatively unhelpful), others have suggested "organization," "levels of organization," "hierarchical steps," and so on. We have no strong argument against any of these alternative terms, even if we know not how to define them, let alone quantify them as we feel we must. By contrast, we have been specific and numerical while using "energy rate density" as an empirical estimate, if not true measure, of complexity. Though utterly foreign among practitioneers of the biological sciences, this term remains straightforward, physically intuitive, robust, and inclusive. Free energy rate density is a serious diagnostic tool, yet to repeat a caveat voiced earlier, a more thorough analysis of open systems' emergent properties must account for not just quantity of energy flow, but also the quality and efficiency of that flow. While clearly lacking predictive power, free energy rate density does seem to have widespread explanatory power, indeed like no other we know for such a vast spectrum of all-scale phenomena observed everywhere in Nature.

*What Is the Ultimate Source of Order and Complexity?*

Since non-equilibrium conditions are prerequisite to create and sustain a flow of free energy, cosmic expansion would seem to be a leading candidate for the ultimate source of order and complexity in the Universe. At a most basic level, then, gravitation was a promoter in the evolution of all organization, a statement sure to dismay most biologists. Some might contend that the onset of any complexity owed more to the interactions of systems with their environments—phenomena most obvious for biological evolution, yet as we have seen also applicable to physical and cultural evolution as well. However, such interactions better fit the category of effect than of cause, the cause itself being the global conditions needed to generate those interactions, and before that the universal expansion needed to generate those conditions. Other suggestions for an ultimate source of order include thermodynamics—especially the second law—but only when that law operates in non-equilibrium, dynamical settings. In any case, our origins date back to the origins of free energy—most recently from the Sun to power life on Earth, before that from the gravitational infall of the proto-Sun that gave rise to nuclear fusion, and before that from the disequilibrium begat in the early Universe. Simply and solely stated, if complexity is the effect, then universal expansion is the cause.

*Is Cosmic Evolution Anthropocentric?*

Our analysis of free energy rate density flowing through open systems has indeed rendered humankind—its brains, its society, and its inventions—near the top of the rising curve of complexity. To some readers, it will come as a surprise that human beings and their institutions process more energy per unit mass than do stars or galaxies, but it should not puzzle us that we are more complex than less-evolved life forms and that they, in turn, are more complex than any inanimate object. We are, quite frankly, a very recent addition to the story of cosmic evolution—a few million years out of 12 billion being well less than a tenth of one percent of the story—and it is not unreasonable that one or another recently ordered system enjoy the topmost entry in the hierarchical array of evolved systems. At this particular epoch in time, and without any evidence for life elsewhere in the Universe, humankind does seem to be that paramount system. These are neither value judgements nor claims about "success" or "wisdom," nor are they meant to imply any sort of anthropocentrism whatsoever. What is different about humankind—among *all* complex systems—is that we are the only known entity, living or not, that has attained the capability of manipulating matter enough to potentially enter the Life Era.

## Is Life Special among Organized Systems?

Life does benefit from what we have called "value-added," adaptive qualities—neo-Darwinian-aided complexity enhanced beyond those strictly energy-driven complexities characterizing most ordered, inanimate objects. Life must function as well as be structured, and it must reproduce lest it have to originate from scratch each time. It's as though biologically derived complexity, fostered by a more subtle and elegant Darwinism, is superposed on the more vigorous and physically derived complexity engendered by non-equilibrium conditions—much as thermodynamic analysis might well explain the structural and even metabolic aspects of life, yet perhaps fall short of guiding its functional and reproductive properties. Even so, we assert that life differs not in kind from non-life, rather only in degree—in fact, degree of complexity—and therefore, however humbling, does not deserve the adjective "special." Not even natural selection is unique to life; selection is more prevalent for living systems because their complexity and diversity are greater, giving rise to myriad opportunities for selection to act, but it can be seen to operate among non-life as well. Nor is the holistic notion of the "whole exceeding the sum of its parts" a property solely of biological systems; a demonstrably inanimate watch clearly qualifies (for its springs, gears, and crystals, or these days its silicon gadgetry, can actually tell us the time!), *and,* like life, a watch needs a flow of energy both to build it and to keep it working. In short, no one has ever discovered anything akin to an *élan vital,* or special life force, that would truly set aside life from all other organized systems.

## Is Cosmic Evolution Reductionistic?

Many people dislike numbers; it's mostly a consequence of the poor teaching of science and mathematics at all levels of today's society. Even among scientists, we often have a disconnect; the biological community generally abhors equations, fearing that the physical sciences are seeking a hostile takeover. Yet, quantitative details and mathematics in particular merely represent a language, a compact way of presenting hypotheses and the results of tests of those hypotheses. The fact that some equations and many numbers are liberally spread throughout this book does not mean that this is a study in reductionistic science. Far from it. I have openly used the word "holistic" probably to the chagrin of many colleagues, all the while realizing that, like the twin nodes of chance and necessity, both synthesis and reductionism are vital aspects of any attempt to decipher the bigger picture as well as the fine details on which that picture is based. In the spirit of one best appreciating a cake *both* by baking (synthesizing) it and by eating (reducing) it, perhaps a better term to describe our effort would be "constructionistic."

### Is Biology Part of Physics?

Clearly, this is a loaded question that bothers biologists no end. It does seem that the physics of irreversible processes might someday broaden to encompass much of biology, and it is also true that physical law restricts avenues of development and evolution available to biological systems. However, it is one thing to say that the phenomena of biology are consistent with the principles of physics; it is quite another to claim that those principles will prove sufficient to fully explain biology. It is certainly untrue that biology is the subject of chance and physics the subject of determinism; all sciences share these twin concepts that play such vital roles in cosmic evolution. Given the idea that neo-Darwinian-derived complexities might be superposed on those more conventionally generated by energy alone, perhaps we can agree that biology is physics with added features. That may be why, in physics, the simplest model for any particular phenomenon is usually the right one, whereas in biology historical accidents and accumulating complexity often invalidate the influence of Ockham's razor. That added degree of complexity might also explain why biology has no natural laws per se, only a set of coarse guidelines. Hence, it remains unclear if *all* evolutionary events embody inherent properties of matter traceable to unifying physical laws. We did not set out, nor have we managed, to reduce evolutionary biology to thermodynamical physics, but if non-equilibrium science can deepen biology's physical foundation, then so much the better.

### Is Cosmic Evolution Progressive?

Humankind is unquestionably making progress in its understanding of Nature writ large, accumulating pertinent, quality information at a feverish pace, and raw bits and bytes even faster than that. The resulting advances are dramatically revealing the nature of the Universe as well as raising scores of new (and good) questions about it. Flat-footedly stated, we do know more than our forebears. Post-modernist shrieks notwithstanding, we are clearly moving toward a definite goal of learning more about the Universe and its multitudinous components, and it is in this sense that humankind is progressing in our quest for knowledge. Cosmic evolution, to make a pun, is quite literally a work in progress. This is not to say, however, that Nature itself makes progress; there is no known specific goal or ultimate stage that the Universe aspires to achieve. Nor is there compelling logic or evidence whatsoever to support the notion that evolutionary events and processes generate progress or are in any way progressive. That said, purists might argue that since we are an integral part of the Universe, it, too, must perforce make progress. Progress is a relative word in a contingent cosmos.

*Where Is the Rising Curve of Complexity Headed?*
No one knows—or can know—where the curve of rising complexity is specifically headed, but if Nature's progress is moot, then the general answer is nowhere in particular other than, presumably, toward richer states of ever-greater complexity. With stochasticity ubiquitous in a changing Universe, let alone the specter of chaos looming large, modern science is unable to predict outcomes of evolutionary events. Not even physics, traditionally the most precise of the sciences, is in the prediction business any more. Statistical models and computer simulations can be devised to suggest potential scenarios for future change, and sometimes accurate forecasts can be made for localized, specialized, and short-duration events, but with chance operative at bifurcation points wherever energy flows break symmetries, long-term products are invariably indeterminate. Even aided by the most powerful computers, we may never be able to predict which of all possible systems will arise and evolve under the action of any given energy flow. A perpetual stream toward richness, diversity, and complexity, the outcome of which cannot be foreseen, may be the true fate of the Universe.

*Why Is Evolution so Controversial?*
Time, life, and complexity are all intricately engaged in the ongoing quest to decipher evolution, yet each is nearly impossible to define in a way on which good and reasonable people, even among experts, can agree. These three terms comprise some of the most enigmatic aspects of our natural world and controversy inevitably results whenever science addresses them, especially in the context of our own selves and not

merely of a rocky planet or a distant star. Together, this trio of terms is at the core of a cosmic-evolutionary scenario that attempts to solve the time-honored mystery of who we are and whence we came—and that is what the controversy is mostly about: ourselves. Despite all the stars and galaxies that form a backdrop, cosmic evolution is a story that places life and humanity on center stage—and that's not an anthropocentric sentiment as much as an honest statement about human curiosity and inventiveness. To be sure, curiosity is the essential driver in our pursuit to know more and it is ultimately the satisfaction of curiosity that will diminish the controversy.

### Is There an Underlying "Platonism" in the Universe?

The question can be effectively rephrased: Are Nature's many varied and oddly diverse structured systems the result of hopelessly entangled causes and effects that are virtually unsolvable, or are those systems governed by unifying, unchanging principles incorporating both randomness and determinism at work behind the temporal flux of all events in the Universe? No one knows, of course, but the second law of (non-equilibrium) thermodynamics, especially its guidance of ubiquitous energy flows, does seem as good a candidate as any principles yet found. The search will continue for commonalities and regularities in Nature, for underlying, recurring order pointing the way toward the final dream of a complete unification of all science, for broad and universal principles that include Nature's countless experiences. Decidedly, modern science, with its physical laws and symmetry rules expressed in mathematical language, is not inconsistent with the Platonic ideal of a deeper reality of "eternal ideas and unchanging forms." Perhaps the real question is this: Are we creating an intellectual map or merely uncovering one that already exists? Such a question is much akin to another age-old mystery: Is mathematics an invention of the human mind or are mathematicians discovering a universal math independent of their interpretations?

### What Does Cosmic Evolution Imply about Extraterrestrial Life?

At the least, cosmic evolution is a story about ourselves—our origin, maturity, and relationship to all known material things. At most, it says that life elsewhere is distinctly possible, perhaps even probable, for the route to life on Earth seems to have been in no way special. Although time is the enemy of all things made, much time naturally gives rise to the cosmic conditions conducive to the emergence of significant complexity. And, if the realization of potential order is the norm at localized sites within an expanding Universe, then increasing complexity might be *expected* to arise at widespread places beyond Earth. But does rising complexity alone guarantee the emergence of life, or might there be a whole family of routes from simplicity to complexity, only one or a few of which inevitably lead to life? Contrary to popular opinion, cosmic evolution makes no requirement that life exist elsewhere; even if we are alone in the Universe, cosmic evolution is still a valid hypothesis of the myriad changes that led so remarkably to us.

*What Is the Status of the Life Era?*
Two extremes dominate: If extraterrestrial life does exist, and if that life is more advanced (i.e., more complex) than we, then humankind is merely passing into an already established Life Era; such an era of active material manipulation would have been fashioned by those who preceded us along the arrow of time. But if we are alone as intelligent beings in a largely infertile cosmos, then we are *creating* that Life Era; we would also be the sole proprietors of the knowledge base discussed in this book, implying perhaps that we have a moral obligation to survive—a responsibility to see to it that this grand experiment called intelligent life does not end. Yet the truth may lie midway: Given the time it took for the cosmic conditions to ripen, the energy to flow, and complexity to grow, we are perhaps among the vanguard of many newly emerging intelligences scattered throughout the expanding Universe, in effect a garden-like cosmos about to flower with uncountable advanced life forms all having originated only relatively recently and all only now beginning to populate the Life Era.

*What Are the Implications of Cosmic Evolution for Religion?*
Make of it what you will; this is a book of science. The evolutionary epic told here is as ennobling as any religion—enlightening, majestic, awesome, providing a sense of the "ultimate." Material reality, when scientifically analyzed in both depth and breadth, brings to mind not only elegant grandeur and a sacred narrative comparable to any religious tradition, but also enriching empiricism and a genuine connection to the cosmos extending into deep history much older than most religions. Such is our grand synthesis at the turn of the millennium, not a replacement for religion as much as a scientific philosophy in its own right, combining testable ideas, penetrating observations, and veritable

inspiration while trekking along that everlasting path toward heightened understanding.

Modern scientific research helps us realize that we are connected to distant space and time not only by our imaginations but also through a common, cosmic heritage: Most of the chemical elements comprising our bodies were created billions of years ago in the hot interiors of remote and long-vanished stars—a physical, stellar metabolism, no less. Their hydrogen and helium fuel finally spent, these giant stars met death in cataclysmic supernova explosions, scattering afar the atoms of heavy elements fused deep within their cores. Resembling a "galactic ecosystem" whose interrelated components are as rich and diverse (though not as complex) as those of life in a tidepool or a tropical forest, this loose interstellar matter eventually collected into huge gas clouds which, in turn, slowly contracted to give birth to a new generation of stars, among them the Sun and its family of planets nearly five billion years ago. Drawing upon the matter gathered from the debris of its stellar ancestors, planet Earth then provided the conditions that eventually gave rise to life and intelligence, and ultimately to ourselves—a biological and cultural metabolism, no more. Like every object in our Solar System, every living creature on Earth embodies atoms from distant realms of our Galaxy and from a past far more remote than the beginnings of human evolution.

A truly long time has passed since the root cause of all order emerged as the matter and radiation fields decoupled in the early Universe. Gradients forevermore having been enabled by the expanding cosmos, it was and is the resultant flow of energy among innumerable non-equilibrium environments that triggered, and in untold cases still maintain, ordered, complex systems on domains large and small, past and present. This general tendency to organize among systems experiencing optimal energy flows is the hallmark of evolution broadly conceived, a potent, recurrent phenomenon widely perceived to be at work throughout Nature in the aggregate—the crux of an underlying, unifying fabric of science like no other. To be sure, the onset and evolution of galaxies and stars and planets and life, indeed even the cultural and social development of humanity, all have rich historical precursors reaching back to events of a much earlier period in natural cosmic history. Consummate products to date, we beings have now become sentient life forms able not only to reflect upon the awesome evolutionary process that brought us forth, but also to attempt to understand it, adopt it, and embrace it.

This book-length essay has addressed many things, few in detail and rigor, most of them broadly and briefly. It has:

- stressed the idea that Nature is not clean and clear, not simple and equilibrated, not "black and white," but rather locally complex with shades of grey throughout; chance mixes with necessity, reductionism with holism, physics with biology;

- synthesized a chronological narrative based on mainstream science, in fact not much more than energy budgets of structured systems, without appealing to any manifestly new science;

- suggested a universal framework, an expanding cosmic environment, in which to decipher the growth of order and complexity among all natural entities;

- proposed a novel way to integrate life—not necessarily just terrestrial life, and not merely with words and sentiments alone—into a sweeping cosmological perspective;

- developed a reasonable claim for a whole new era—the Life Era—as part of a rich and comprehensive natural history of the Universe;

- reconciled, in general terms, the evident constructiveness of cosmic evolution with the inherent destructiveness of essential thermodynamics;

- offered a bold and coherent worldview, an inclusive evolutionary epic about the Universe and ourselves that virtually everyone can comprehend.

Cosmic evolution propounds a unifying synthesis to use as a grand ethos of potentially unprecedented intellectual magnitude while approaching an uncertain future. Not handed to us on a stone tablet, and hence itself subject to change as research progresses, this pervasive worldview nonetheless embodies a powerful guide to an essential understanding of the nature of all material things. Looking backward toward the past, we sense that its central theme—the time-honored concept of change—can account for the appearance of matter from the primal energy of the Universe, and in turn for the emergence of life from that matter. Looking forward toward the future, we search for enhanced understanding, for meaning and rationality. Just how wise, quite aside from sheer intelligence, are we? Is humankind part of a cosmological imperative, heading, perhaps with other sentient beings, toward some astronomical destiny? Put bluntly yet magnanimously, the scenario of cosmic evolution grants us unparalleled "big thinking," from which may well emerge the global ethics and planetary citizenship likely needed if our species is to remain part of that same cosmic-evolutionary scenario.

SYMBOLS AND NUMERICAL CONSTANTS
GLOSSARY · NOTES · WORKS CITED
FURTHER READING · INDEX

# SYMBOLS AND NUMERICAL CONSTANTS

| | |
|---|---|
| $a$ | radiation constant = $7.56 \times 10^{-15}$ erg cm$^{-3}$ K$^{-4}$ |
| $B$ | number of possible structural arrangements |
| $B_\lambda$ | radiation intensity |
| $c$ | velocity of light = $3 \times 10^{10}$ cm s$^{-1}$ |
| $C$ | specific heat capacity |
| $E$ | total energy |
| $F$ | free energy |
| $G$ | gravitational constant = $6.67 \times 10^{-8}$ dyn-cm$^2$ g$^{-2}$ |
| $h$ | Planck's constant = $6.63 \times 10^{-27}$ erg-s |
| $I$ | information |
| $H$ | Hubble's constant |
| $H_0$ | Hubble's constant today |
| $k$ | spacetime curvature factor |
| $k_B$ | Boltzmann's constant = $1.38 \times 10^{-16}$ erg K$^{-1}$ |
| $K$ | dimensionless unit of information content = 1 bit |
| $L$ | luminosity |
| $m$ | mass of a particle |
| $M$ | mass of a system |
| $n_\lambda$ | number density of photons in a wavelength range centered on $\lambda$ |
| $N$ | number of gas particles |
| $p$ | probability |
| $Q$ | heat energy |
| $r$ | radius or distance |
| $R$ | universal scale factor |
| $R_g$ | ideal gas constant = $8.3 \times 10^7$ erg mole$^{-1}$ K$^{-1}$ |
| $S$ | entropy |
| $S_{sys}$ | entropy of a system |
| $S_{env}$ | entropy of an environment |
| $t$ | time |

| | |
|---|---|
| $T$ | temperature |
| $T_m$ | temperature of matter |
| $T_r$ | temperature of radiation |
| $u$ | energy density of radiation |
| $U$ | gravitational potential energy |
| $v$ | velocity |
| $V$ | volume |
| $W$ | number of microscopic states |
| $z$ | Doppler shift |
| $\delta$ | symbol for change |
| $\Phi_m$ | free energy rate density |
| $\lambda$ | wavelength |
| $\Lambda$ | cosmological constant |
| $\rho$ | density |
| $\rho_m$ | mass density |
| $\rho_{m,c}$ | critical mass density |
| $\rho_{m,0}$ | current mass density |
| $\rho_r$ | equivalent mass density of radiation |
| $\sigma$ | Stefan-Boltzmann constant $= 5.7 \times 10^{-5}$ erg cm$^{-2}$ s$^{-1}$ K$^{-4}$ |
| $\Omega$ | universal density parameter |

# GLOSSARY

**adaptation**   The response to a changing environment of an organism's structure or function in a way that improves its chance for survival and reproduction.

**amino acid**   An organic molecule containing carboxyl and amino groups, of which 20 different types form the building blocks of the proteins that direct the metabolism in all life forms on Earth.

**anthropic principle**   The idea that the Universe is the way it is because we (intelligent beings) are here to observe it; the Universe is made for us.

**anthropocentrism**   The idea that events can be viewed and interpreted in terms of human activities and values.

**anthropology**   The study of humanity, including its origins, physical development, culture, race, social customs, and beliefs.

**arrow of time**   In thermodynamics: the irreversible and inexorable increase in entropy for all natural events. In cosmology: the regular and apparent increase in complexity throughout the history of the Universe.

**astronomy**   The study of material events in the Universe beyond Earth's atmosphere.

**astrophysics**   The study of the interaction between matter and radiation in space.

**atom**   A submicroscopic component of matter, composed of positively charged protons and neutral neutrons in the nucleus, surrounded by negatively charged electrons.

**atom epoch**   A period in the early Universe when elementary particles began to cluster, thus fashioning the first atoms.

**ATP**   An acronym for adenosine triphosphate, an organic molecule that acts as energy currency in life forms; the central conveyor of phosphate-bond energy in a cell's metabolism.

**autocatalysis**   The acceleration of a chemical reaction when the product of that reaction is also a catalyst for the same reaction.

**autopoesis**   A quality of systems having the power to generate themselves, meaning literally "self-creation"; in biology, the idea that life is best characterized not only by structure and function but also by process, which is more important than interaction with the environment.

**autotroph** Any organism capable of meeting its nutritive needs by feeding on inorganic matter and external energy, such as plants with the help of sunlight.

**baryons** Matter composed mainly of protons and neutrons; "normal" matter (as opposed to "dark" matter) comprising stars, planets, and life forms.

**big bang** A popular term describing the explosive start of the Universe.

**bifurcation** The division of a phenomenon into two solutions while a system parameter (energy, in this book) varies with time.

**biochemistry** The study of chemical processes in living organisms.

**biological evolution** The changes experienced by life forms, from generation to generation, throughout the history of life on Earth.

**biology** The study of life in all its forms and phenomena.

**biosphere** That part of Earth's crust, water, and atmosphere where living organisms exist and undergo their life cycles.

**catalyst** A facilitator or accelerator of a chemical reaction without itself being consumed or changed in the process.

**cell** A minimal, usually microscopic, chemical system that can be recognized as alive.

**central dogma** In biology: the assertion that information in biological systems passes unilaterally from nucleic acids to proteins, but not conversely. In physics: the assertion that energy is conserved in all systems, in all environments, and at all times in the Universe.

**chance** A happening without known cause; fortuitous, accidental, contingent.

**change** To make different the form, nature, and content of something; the transformation of one system into another that is different in at least one respect.

**chaos** In the old sense, unconstrained randomness, disorder; in the new sense, the behavior of a deterministic system under conditions that allow for the possibility of multiple outcomes.

**chemical evolution** The pre-biological changes that transformed simple atoms and molecules into the more complex chemicals needed for the origin of life.

**chemistry** The study of the properties, compositions, and structures of substances and elements, and the ways they interact with one another.

**chemosynthesis** The production of organic matter by microorganisms that use chemical energy stored in certain inorganic substances, such as hydrogen sulfide.

**classical physics** A branch of physics dealing mostly with deterministic mechanisms; the worldview according to Newton.

**closed system** A system able to exchange energy, but not matter, with its surrounding environment.

**complexity** A state of intricacy, complication, variety, or involvement, as in the interconnected parts of a structure—a quality of having many interacting, different components; operationally, a measure of the information needed to describe a system's structure and function, or of the rate of energy flowing through a system of given mass.

**consciousness**   That property of human nature generally, or of the brain specifically, that grants us self-awareness and a sense of wonder.

**conservation of mass and energy**   A basic principle of science stipulating that the sum of all mass and energy in a closed system remains constant during any event.

**convection**   The transfer of heat via circulation, resulting from the upwelling of warm matter and the concurrent downward flow of cool matter to take its place.

**cosmic background radiation**   A weak, nearly isotropic electromagnetic (mostly microwave) signal permeating all of space, thought to be a remnant of the big bang.

**cosmic evolution**   The sum total of all the many varied changes in the assembly and composition of radiation, matter, and life throughout the history of the Universe.

**cosmological principle**   A basic assumption of modern cosmology, namely that the Universe is homogeneous (uniform at every point) and isotropic (uniform in every direction) on scales larger than galaxy superclusters.

**cosmology**   The study of the structure, evolution, and destiny of the Universe.

**cosmos**   A complete, orderly, harmonious system; from the Greek, *kosmos,* meaning an orderly whole.

**creation**   An act of producing or causing to exist.

**culture**   The totality of artifacts, behavior patterns, institutions, and mental constructions acquired by members of society through learning.

**cultural evolution**   The changes in the ways, means, actions, and ideas of societies, including the transmission of same from one generation to another.

**dark matter**   Unseen mass in galaxies and galaxy clusters whose existence is only inferred indirectly, but which has not been confirmed directly by any observations.

**Darwinism**   The idea that living species originate by descent with variation from parental forms, by means of natural selection of those best adapted to survive in the struggle for existence.

**decoupling**   An event in the early Universe when atoms first formed, after which photons moved freely in space, causing matter and radiation to behave differently.

**density**   A measure of compactness, namely the quantity of something in a unit of volume.

**determinism**   The idea that all events have specific, definite causes and obey precise, natural laws, making their outcomes completely predictable; from any particular initial state, one and only one sequence of future states is possible.

**development**   Any process of change, usually of growth or elaboration, between a system's origin and its maturity.

**disorder**   An irregularity in arrangement or behavior; a synonym for entropy; an absence of order.

**dissipative structure**   A dynamic system capable of dispersing both energy and entropy under non-equilibrium conditions.

**DNA**   An acronym for deoxyribonucleic acid, a self-replicating molecule resident chiefly in biological nuclei, mainly responsible for transmitting hereditary information and for the building of proteins.

**Earth** Humankind's home planet in space, third out from the Sun.

**ecology** The study of the interrelatedness among all systems, and between those systems and their environment; most common in biology, in which the systems comprise all living things.

**ecosystem** A community of systems and their shared environment, regarded as a unit, all interacting so as to perpetuate the grouping more or less indefinitely; most common in biology, in which the systems are plants, animals, and other organisms, and the environments are often seafloor, forest, and grassland areas.

**element** A substance comprising one and only one distinct kind of atom; one impossible to separate into simpler substances by chemical means.

**emergence** The appearance of entirely new system properties at higher levels of complexity not pre-existing among, nor predictable from knowledge of, lower-level components; the process of a system "becoming" from its environment at certain critical stages in its development or evolution.

**energy** The ability to do work or to produce change; an abstract concept invented by nineteenth-century physicists to quantify many different phenomena in Nature.

**energy density** A measure of compactness of energy; an amount of energy per given volume.

**entropy** A measure of randomness, or disorder, of a system, reaching a maximum at thermodynamic equilibrium; a lack of information about a system's organization.

**environment** Any part of the Universe not included in a system; a combination of all things, conditions, and influences surrounding a system.

**enzyme** Any of numerous complex proteins that catalyze specific biochemical reactions.

**equilibrium** A state wherein a system's gradients are negligible, its probability maximized and its free energy minimized; one that constantly reacquires any and all of its possible configurations randomly, thus from which it exhibits no tendency to depart.

**eukaryote** A life form whose cells have well-developed biological nuclei; all organisms above the level of prokaryotes, including protists, fungi, plants, and animals.

**event** Any occurrence in spacetime; a happening.

**evolution** Any process of formation, growth, and change with time, including an accumulation of historical information; in its broadest sense, both developmental and generative change.

**first law of thermodynamics** A principle stipulating that, in any real process, energy is conserved, that is, never created or destroyed but allowably changed from one form to another.

**flow** The movement of an entity from one place to another; to issue or proceed from a source.

**force** An agent of change in or on any system.

**form**   The structure, pattern, organization, or essential makeup of anything.

**free energy**   That type of energy available to do useful work; a measure of the amount of change possible in a system.

**free energy rate density**   The amount of energy (available to do work) flowing through a system per unit time and per unit mass.

**function**   The ability of a system's components, beyond its mere structure, to execute an internal action or role, such as breathing, running, writing, or reproducing.

**galactic evolution**   The changes experienced by galaxies, either intrinsically because of localized changes among myriad stars, or environmentally because of merges, acquisitions, and close encounters among neighboring galaxies.

**galaxy**   An open, coherent, spacetime structure maintained far from thermodynamic equilibrium by a flow of energy through it—a colossal system of billions of stars and loose gas held together by gravity.

**galaxy epoch**   A period in the relatively early history of the Universe when the galaxies formed.

**gene**   A segment of any DNA molecule containing information for the construction of one protein, hence responsible for directing inheritance from generation to generation.

**genetics**   The study of heredity and the biological processes by which inherited characteristics are passed from one generation to the next.

**geology**   The study of the physical history of Earth, especially the rocks of which it is composed and the processes it has undergone.

**gravity**   An attractive force that any massive object exerts on all other massive objects.

**hadron epoch**   A very early time in the history of the Universe when heavy, strongly interacting, elementary particles, such as protons and neutrons, were the most abundant type of matter.

**heat**   The amount of energy transferred to or from a substance; the thermodynamic state of an object by virtue of the random motions of the particles within it.

**heredity**   The transmission of genetic traits from parents to offspring, thus ensuring the preservation of certain characteristics among future generations of a species.

**heterotroph**   Any organism requiring organic matter for food, such as primitive cells that survived by absorbing acids and bases floating on primordial seas.

**holism**   The idea that a whole entity, as a basic component of reality, has an existence greater than the mere sum of its parts.

**information**   The number of bits needed to specify a message or structure; the difference between the maximum possible and actual entropies of any given system.

**intelligence**   The capacity to comprehend relationships; a biological adaptation for complex behavior, probably synonymous with language.

**irreversible process**   An event occurring in only one direction, thereby resembling a non-equilibrium state.

**isolated system**  A system totally separated from its surrounding environment, thus unable to exchange either matter or energy.

**kinetic energy**  The energy of an object or system due to its mass and motion; the ability to do work actively via motion.

**Lamarckism**  The idea that an organism is a result of environmental influences rather than genetic inheritance; traits can be acquired through habit, use, or disuse during a single lifetime and then passed on intact to the next generation.

**lepton epoch**  A very early time in the history of the Universe when the lightweight, weakly interacting, elementary particles, such as electrons and neutrinos, were the most abundant type of matter.

**life**  An open, coherent, spacetime structure maintained far from thermodynamic equilibrium by a flow of energy through it—a carbon-based system operating in a water-based medium, with higher forms metabolizing oxygen.

**Life Era**  A period in the history of the Universe (just beginning on Earth) when technological life forms manipulate their genes and their environments more than conversely.

**mass**  A measure of the total amount of matter, or "stuff," contained within an object.

**materialism**  The idea that there is nothing other than matter, energy, and their various arrangements and motions in the Universe.

**matter**  Anything that occupies space and has mass.

**Matter Era**  A period in the history of the Universe (including now) when the density of energy contained within matter exceeded the density of energy contained within radiation.

**mechanism**  The idea that all natural processes are machines, explainable in terms of Newtonian mechanics and thus ultimately predictable.

**metabolism**  The sum of all chemical reactions that energetically support a living organism, starting from energy sources that are either chemical (environmental nutrients) or physical (sunlight).

**Milky Way**  Humankind's home galaxy, comprising some hundred billion star systems, so named because its stars resemble a milky band running across the dark night sky.

**modern synthesis**  A conceptual unity in contemporary biology, based on Darwinian evolution, including natural selection, adaptation, diversity, and Mendelian genetics; also termed neo-Darwinism.

**molecule**  A bound cluster of two or more atoms held together by electromagnetic forces.

**mutation**  A random, microscopic change in one or more genes of an organism, transmissible by replication.

**natural selection**  In general, a normative process whereby environmental resistance tends to eliminate non-randomly those members of a group of systems least well adapted to cope and thus, in effect, choose those best suited for survival. In biology, the Darwinian process whereby a population's life forms having advantageous traits are able to adapt to a changing environment, thereby surviving, reproducing, and passing on to their descendants those favorable traits, which then accumulate in the population over time.

**Nature**   The Universe, including all its natural phenomena.

**negentropy**   A measure of entropy whose value, in this case, is less than zero, i.e., negative entropy, or orderliness, information-rich.

**neo-Darwinism**   A combination of traditional Darwinism and Mendelian genetics; also termed "the modern synthesis."

**non-equilibrium**   A state characterized by non-negligible gradients and a regular energy flow, allowing for further change, growth, and evolution.

**nucleic acid**   A class of long-chain, organic molecules, made by grouping many nucleotides and often inhabiting the biological nuclei of cells.

**nucleotide base**   An organic molecule, of which 5 different types comprise the building blocks of all nucleic acids within genes that transmit hereditary characteristics from one generation of life forms to the next.

**ontogeny**   The developmental history of an individual system; in biology, the developmental life cycle of an individual organism.

**open system**   A system able to exchange both energy and matter with its surrounding environment.

**order**   A regularity in arrangement or behavior; a restriction on the number of possible states; an absence of disorder.

**organism**   Anything that lives—plant, animal, or microbe—or has ever been living.

**organization**   Relations existing among the components of a system for it to be a member of a specific class.

**origin**   A coming into being; a process whereby a given state precedes all other such states in time.

**parsec**   The distance to an object subtending a half-Earth-orbit parallax of exactly 1 arc second; equal to $3.1 \times 10^{18}$ cm, or 3.26 light-years.

**particulate evolution**   The changes among elementary particles, including photons, in the early Universe.

**photon**   A packet of pure energy; the massless, chargeless carrier of the quantum of electromagnetic radiation.

**photosynthesis**   The production of organic matter by (usually) green plants that use sunlight to make glucose from carbon dioxide and water, the byproduct being oxygen.

**phylogeny**   The evolutionary history of a group of systems; in biology, the evolution of all species or of a particular group of species.

**physics**   The study of matter, energy, space, and time.

**planet**   An open, coherent, spacetime structure maintained far from thermodynamic equilibrium by a flow of energy through it—a rocky and/or gaseous system, more massive than an asteroid yet less massive than the star about which it orbits.

**planetary evolution**   The changes in the physical or chemical properties of planets during the course of their histories.

**Platonism** The philosophy that the changing, shifting world of physical phenomena masks a deeper reality—an underlying set of eternal ideas and unchanging forms, and it is these alone that grant true knowledge.

**potential energy** The energy of an object or system due to its mass and position; the ability to do work passively stored.

**power** The rate at which work is done; an amount of energy transferred per unit time.

**probability** The likelihood of an event, expressed by the ratio of the number of actual occurrences to that of possible occurrences.

**process** The change of a quantity over time; the act of proceeding.

**prokaryote** A life form whose single cell lacks a well-developed biological nucleus, such as various types of bacterial microorganisms.

**protein** A class of large, organic molecules, made of many amino acids and inhabiting the cytoplasm of cells; a major structural component of all animal tissues, and a functional enzyme in both plants and animals.

**punctuated equilibrium** The idea that life's species remain essentially unchanged for long periods of time, after which they change rapidly in response to sudden, drastic changes in the environment.

**quantum physics** A branch of physics dealing with sub-microscopic parts of a system, including its inherent uncertainty.

**radiation** A form of energy that travels at the velocity of light, of which light itself is a special kind.

**Radiation Era** A period in the history of the Universe (early on) when the density of energy contained within radiation exceeded the density of energy contained within matter.

**reductionism** The idea that all natural phenomena can be understood only by reducing them to their smallest component parts.

**relativistic physics** A branch of physics dealing with matter moving at high speeds (special theory) or in strong gravitational fields (general theory); the worldview according to Einstein.

**reversible process** An event able to proceed in either a forward or a reverse direction, thereby resembling an equilibrium state.

**RNA** An acronym for ribonucleic acid, a single-stranded organic helix found chiefly in the cytoplasm of cells, often instrumental in protein synthesis.

**second law of thermodynamics** A principle stipulating that, in any real process, the entropy of the Universe increases, that is, irreversibly tends toward greater disorder; energy naturally flows from hotter to colder systems, and not in the reverse.

**selection** A process of Nature that causes some systems having certain properties, which are not the norm for a population of systems, to preferentially adapt to their environment and thus to enhance their state; those things that work well survive, and those that don't, don't.

**self-organization** A phenomenon whereby an ordered entity emerges partly on its own (when out of equilibrium) and partly aided by the flow of energy (at a bifurcation point), often displaying increased order and complexity.

**simplicity** A state free of complexity or of the possibility of confusion.

**Solar System** Humankind's home planetary system, comprising nine planets, dozens of moons, and countless smaller asteroids and meteoroids, all orbiting the Sun.

**space** An indefinitely great three-dimensional expanse in which all material objects are located and all events occur.

**species** Any organism—plant, animal, or microbe—of a single kind; a fundamental biological classification denoting not only individuals that are structurally similar but also those able to mate among themselves and produce fertile offspring.

**standard model** In physics, an acknowledged description of microscopic phenomena, bolstered by accelerator experiments and the quantum field theory of particles and forces. In cosmology, an acknowledged description of macroscopic phenomena, bolstered by observations of galaxy recession, background radiation, and elemental abundances.

**star** An open, coherent, spacetime structure maintained far from thermodynamic equilibrium by a flow of energy through it—a round, gaseous system so hot that its core can sustain thermonuclear fusion.

**state** The status or condition of a system as specified by certain dynamical variables.

**statistical physics** A branch of physics dealing with vast numbers of particles and their probable states, enabling some averaging of properties within macroscopic systems.

**statistical fluctuation** Continual change from one course or condition to another; an irregularity or instability.

**stellar epoch** A period in the history of the Universe (including now) when the stars form.

**stellar evolution** The changes experienced by stars as they originate, mature, and terminate.

**structure** The arrangement of the basic components of a system, including form but not function.

**Sun** Humankind's parent star, a resident of the Milky Way.

**symbiosis** The living together, in a mutually beneficial union or close association, of two dissimilar organisms.

**symmetry** The ordered repetition of identical parts of a structure or state.

**system** A finite assemblage of interdependent things in the Universe, separated from its surrounding environment by topological and organizational boundaries; any entity of interest, usually one having interconnected components acting as a unitary whole.

**teleology** The idea that Nature is governed by design or purpose that is extraphysical; the philosophy of final causes that guide phenomena toward certain goals.

**temperature** A measure of the heat of an object, by virtue of the random motions of the particles within it.

**thermodynamics** The study of the macroscopic changes in the energy of a system, for which temperature is a central property, and meaning literally "movement of heat."

time   The fourth dimension that distinguishes past, present, and future; a quantity easily measured yet hard to define.

Universe   The totality of all known or supposed objects and phenomena, formerly existing, now present, or to come, taken as a whole.

# NOTES

## PREFACE

1. The term "cosmic evolution" is used synonymously with "universal change"—the study of myriad alterations in order, form, and complexity on all spatial and temporal scales. Some researchers have called the subject "cosmography," others "cosmogenesis." Still others, because the subject incorporates the study of life within a cosmological context, prefer the terms exobiology, bioastronomy, or astrobiology, all of which have much the same meaning, albeit restricted agenda. We shall employ the term cosmic evolution, for its scope here is meant to be broader and more encompassing than any of the alternatives.

## PROLOGUE

1. The proposed sequence of ordering—galaxies, stars, planets, and life—holds as a general statement. Galaxies and stars might well have emerged simultaneously ("coevolved"), to be sure some of the oldest globular star clusters (~12 billion years) are taken to be virtually as old as their parent galaxies, none of which, in turn, are much younger than the Universe per se. Conceivably, some stars might have preceded the galaxies, possibly becoming the building blocks for those galaxies; we shall later consider the hypothesized "first stars," and especially their thermodynamic implications for reheating the Universe. But in the main, since galaxies stopped forming (for reasons unknown) in the relatively early Universe, and since stars have continued to form ever since, the epoch of galaxy formation must have mostly preceded the epoch of star formation; the stars we see today almost surely originated only after the hot, gaseous galaxies had time to cool. Likewise, some planet-like assemblages could have conceivably formed prior to the stars, but it's not likely; rather, protoplanets usually emerge contemporaneously with stars, a mature planetary system a little later, followed sometimes (or at least once) by life.

2. The unifying law of Nature is likely to be physical, the underlying pattern conceivably Platonic, and the ongoing process almost certainly energetic (in part). As for a "principle of cosmic selection," one has been already proposed for civilizations broadly conceived: "Those technological civilizations anywhere in the Universe that recognize the need for, develop in time, and fully embrace global ethics will survive, and those that do not will not" (Chaisson 1988, 1999). However, ethics, morality, and the fate of civilizations are topics not well suited for this brief book primarily concerning science.

## INTRODUCTION

1. As the twentieth century's foremost champion of Darwin and a co-founder of biology's modern synthesis, or neo-Darwinism, Ernst Mayr (personal communication 1995) has a way of phrasing the hackneyed "survival of the fittest" that incorporates nicely both chance and determinism. He summarizes Darwin's principle of natural selection in a twofold manner: (1) random production of variations within a population (that's the chance part), and (2) non-random elimination of the less fit (that's the deterministic selection factor).

2. More generally, assuming that n microstates are not equally probable but have individual probabilities $p_i$, the entropy $S = -k \,^{n}\Sigma_{i=1}\, p_i \ln p_i$, where the $p_i$ are positive numbers that sum to 1. A similar generalization pertains to the equation for information content, a few sections hence. Note also that when equal expectations exist, $p_i = 1/W$, and since the unconstrained sum over all configurations yields W total microstates, the general equation above reduces to the simplified form used in the text,

$$S = -k \Sigma W^{-1} \ln W^{-1} = -k \Sigma W^{-1} (\ln 1 - \ln W) = -k \, WW^{-1} (-\ln W) = k \ln W.$$

## 1. MATTER

1. People of many persuasions often judge it appalling when a scientist uses the word "creation." Colleague scientists often demur, assuming capitulation to religious pressures, and theologians become uneasy while thinking, in turn, that scientists are treading on their turf. On the contrary, this book has no problem with the word, indeed uses it often. Whether the origin of the Universe occurred at the hand of a deistic or theistic God beyond our known, material Universe or as part of a completely naturalistic fluctuation in spacetime (see last two paragraphs of note 3 in Chapter Two), neither scientists nor theologians can say with any degree of certainty. But the word "creation" should not be the exclusive purview of religious thinkers.

2. The range of possible error in the age of the Universe is considerable. For two reasons, Hubble's constant is known to an accuracy of no better than 25%. First, although astronomers can measure spectroscopically accurate velocities of galaxies near and far, the rate of galaxy recession depends critically on the rather uncertain distances of remote and thus faint galaxies, which even the largest telescopes have

trouble observing well. Second, the empirically derived value of Hubble's constant is skewed by an uncertain movement of the Local Group of galaxies (of which the Milky Way is one of nearly three dozen members) toward the great Virgo Galaxy Cluster some 50 million light-years away; this net drift amounts to several hundred km s$^{-1}$ and might be nothing more than the random motion of our Local Group in the outskirts of the larger local supercluster of galaxies (of which Virgo and its 25,000 member galaxies probably comprise the core). Even so, and although an error of several billion years seems large, the difference between the extreme possible ages—10 or 20 billion years—is only a factor of two, which is really quite good for an order-of-magnitude subject like cosmology. One of the principal goals of the *Hubble Space Telescope* project, appropriately enough, is a multi-year measurement of Hubble's constant and hence the age of the Universe to an accuracy of 10%, a task that has thus far eluded this mission even after repairs were made to compensate for the telescope's famously flawed mirror.

3. In 1998, observations of supernovae at unprecedented distances roughly halfway back to the beginning of time—~6 billion light-years away—resurrected serious talk about the cosmological constant for the third time in the twentieth century. The so-called $\Lambda$ term might be needed to explain these observations since the supernovae appear dimmer than expected, suggesting that the Universe is accelerating! Apparent brightness of supernovae are taken to be measures of their distances, and therefore of the rate at which cosmic expansion has swept them away over billions of years. Since the supernovae seem less bright, hence farther, than expected, the implication is that the cosmic expansion has sped up since they exploded. This surprising result, if confirmed, would require a mysterious energy, a by-product of $\Lambda$, to permeate the cosmos and boost its expansion rate. The density of the newly hypothesized energy would be constant throughout all space and for all time, so the push it would produce to counter gravity and accelerate expansion is also constant. In the early Universe, when the matter density, $\rho_m$, was larger, gravity would have been strong enough to overwhelm $\Lambda$ and slow the expansion; by contrast, during the past few billion years, $\Lambda$ would have begun to dominate, producing an acceleration as gravity's grip weakened.

As of this writing, the new observations are suggestive, though not compelling. Many other confounding factors, most notably the haze of cosmic dust, could dim the expected light of supernovae; the supernovae of long ago might also be intrinsically dimmer because of their relative lack of heavy elements or for reasons unknown. In any case, the main lines of argument in the cosmic-evolutionary scenario addressed in this book are largely independent of whether the Universe is accelerating or decelerating. Provided that the Universe is expanding, a proposition that is convincing on both theoretical and observational grounds, then the growth of order, flow, and complexity is a natural by-product of cosmic evolution. What's more, if the Universe actually is accelerating, then, as we shall see, it would likely enhance the trend toward richness and diversity among organized systems, for an accelerating Universe would drive environments everywhere further from equilibrium, virtually guaranteeing a perpetual increase in complexity on localized scales, even as the total entropy rises ever more dramatically on universal scales.

4. When v becomes comparable to c, special relativity demands that a relativistic Doppler formula be used,

$$z = v/c \, (1 - v^2/c^2)^{-0.5}.$$

Thus z can become arbitrarily large as $v \to c$, yet reduces to v/c when v<<c.

5. When all the velocity vectors of the Earth, Sun, Galaxy, and Local Group are taken into account, our Galaxy should be moving in the direction of the constellation Aquarius with a net velocity of about 600 km s$^{-1}$ relative to the cosmic background radiation, as discussed in Chapter Two. This agrees with the dipole nature of the observed background radiation, resulting in a minute yet measurable blue shift (2.735 + 0.003 K) toward one region of the sky (Aquarius) and a completely symmetrical red shift (2.735 − 0.003 K) in the opposite direction. These slight anisoptropies (at the 0.1% level) are thus not intrinsic to the cosmic background radiation itself; rather, they are the result of Doppler shifts of that background radiation, allowing us to speak of relative motions against an absolute, stationary frame of reference.

## 2. RADIATION

1. On small spatial scales, the *Cosmic Background Explorer* (COBE) satellite in 1991 found evidence for inhomogeneities in the cosmic background radiation, revealed as variations in blackbody temperature. These variations, impressed intrinsically on the photon distribution at the time of recombination ($\sim$100,000 years) and detected as distortions in the cosmic background radiation in the amount of $\Delta T/T = (5 \pm 1.5) \times 10^{-6}$, probably represent the long-sought thermal seeds for the origin of the galaxies, but an adequate model of such galaxy origins still eludes us. (See also note 9 in this chapter.)

2. Note the uncommon integral, $\int_0^\infty x^3 \, (e^x - 1)^{-1} \, dx = \pi^4/15$.

3. The quest to unify all the known forces of Nature has recently synthesized some aspects of the subjects of cosmology and particle physics. The electromagnetic force binding atoms and molecules and the weak nuclear force governing the decay of radioactive matter have been merged by a theory that asserts them to be different manifestations of one and the same force—an "electroweak" force. Crucial parts of this theory have now been confirmed by experimenters using the world's most powerful particle accelerator, the Conseil Européen pour la Recherche Nucléaire (CERN) near Geneva, and concerted efforts are under way to extend this unified theory to include the strong nuclear force that binds elementary particles within nuclei. Furthermore, though we are unsure at this time how, in turn, to incorporate into this comprehensive theory the fourth known force (gravity), there is reason to suspect that we are nearing the realization of Einstein's dream—understanding all the forces of Nature as different aspects of a single, fundamental force (Weinberg 1994).

The intellectual synthesis of the macrodomain of cosmology (for gravity is a demonstrably long-range force) and the microdomain of particle physics is but a small part of the grand scenario of cosmic evolution. Yet it is an important one, for this newly emerging interdisciplinary specialty of "particle cosmology" could provide great insight into the earliest period of the Universe, the time interval colloquially labeled "chaos."

In brief, descriptive terms, this is the way the newly understood electroweak force operates. In microscopic (quantum) physics, forces between two elementary particles are represented by the exchange of a boson particle; in ordinary electromagnetism familiar to us on Earth, the boson is merely a photon. The electroweak theory includes four such bosons: the usual photon, as well as three other subatomic particles innocuously named $W^+$, $W^-$, and $Z^0$. At $T < 10^{15}$ K—the domain encompassing virtually everything we know about on Earth and in the stars and galaxies—these bosons split into two families: the photon that expresses the usual electromagnetic force and the other three that carry the weak force. But at $T > 10^{15}$ K, these bosons work together in such a way as to make indistinguishable the weak and electromagnetic forces. Thus, by experimentally studying the behavior of this new force, we gain insight into not only the essence of Nature's building blocks but also the early epochs of the Universe, especially the hadron period around $t \simeq 10^{-10}$ s.

To appreciate the nature of matter at $T > 10^{15}$ K and thereby to explore indirectly times even closer to the big bang, physicists are now researching a more general theory that incorporates the electroweak and strong nuclear forces (but not yet gravity). Several versions of this so-called grand unified theory (GUT) have been proposed, though experimentation capable of determining which, if any, of these theories is correct has really only begun. Like the other forces just noted, this grand force is mediated by a so-called X boson elementary particle. It is, according to these grand theories, the very massive (and thus energetic) X bosons that play a vital role in the first instants of time.

At $t = 10^{-39}$ s, when $T \simeq 10^{30}$ K, only one type of force other than gravitation operated: the grand unified force just noted. According to the theory of such a force, the matter of the Universe must have exerted a very high pressure that pushed outward in all directions. The Universe then responded to this pressure by expanding accordingly, and as such $T \sim R^{-1}$ as derived in the main body of the text. Thus, for example, as t advanced from $10^{-39}$ to $10^{-35}$ s, the Universe grew another couple of orders of magnitude and T fell to about $10^{28}$ K.

Now, according to most grand unified theories, a temperature of $10^{28}$ K is special, for at this value a dramatic change occurs in the expansion of the Universe. In short, when $T < 10^{28}$ K, the X bosons can no longer be produced; at $t > 10^{-35}$ s, the energy needed to create such particles was too dispersed owing to the diminishing temperature. As T fell below $10^{28}$ K, the disappearance of the X bosons is theorized to have caused a surge of energy roughly like that released as latent heat when water freezes. After all, energy no longer concentrated enough to yield X bosons was nonetheless available to enhance the general expansion of the Universe, in fact to cause it to expand violently or "burst" for a short duration just after the demise of the bosons. The youthful Universe, though incredibly hot, was quite definitely cooling and in this way experienced a series of such "freezings" while passing progressively toward cooler states of being. Perhaps the most impressive of all such transitions, the rapid decay of X bosons caused a tremendous acceleration in the rate of expansion. This period of (actually exponential) expansion has been popularly termed "inflation"; in a mere $10^{-35}$ s the Universe inflated some $10^{20}$ times or more, smoothing out (by stretching) any irregularities existing at the outset, much as crinkles on a balloon vanish as it is inflated (Guth 1997).

At the conclusion of the inflationary phase at t $\simeq 10^{-35}$ s, the X bosons had disappeared forever, and with them the grand unified force. In its place were the electroweak and strong nuclear forces that operate around us in our more familiar, lower-temperature Universe of today. With these new forces in control (along with gravity), the Universe resumed its more leisurely expansion.

Can we test this GUT proposal, including its implied and spectacular inflationary phase change? The answer is a qualified yes, for we can do so only indirectly. After all, the world's most powerful particle accelerator is only able to create, for the briefest of instants, conditions approximating T = $10^{15}$ K sufficient to confirm the electroweak theory. The grand unified theories become operative at T > $10^{28}$ K, which physicists will likely never be able to simulate on Earth. So, while we have successfully simulated in the laboratory the physical conditions characterizing later parts of the hadron epoch, physicists have little hope of reproducing the earliest chaos period. (To boost subatomic particles to the huge energies needed would require an accelerator machine spanning the distance between Earth and the Alpha Centauri star system some 4 light-years away, and utilizing for a mere few seconds of operation an expenditure of power equal to the cost of several times the gross national product of the United States.)

One especially attractive aspect of the GUTs is that they can generally account for the observed excess of matter over antimatter. It so happens that the decay of the X bosons at t < $10^{-35}$ s lacks symmetry; their decay is expected to have created slightly greater numbers of protons than antiprotons (or electrons than positrons). Specifically, theoretical calculations suggest that for every $10^9$ antiprotons (or positrons), $10^9$ + 1 protons (or electrons) were created; the billion matched pairs subsequently annihilated each other, leaving a $10^{-9}$ particle-to-photon residue of ordinary matter from which all things—including ourselves—have emerged. If this "symmetry-breaking" imbalance is true, then the matter extant today is just a tiny fraction of that formed originally.

This prediction can be tested in a straightforward way, for if protons can be created they can also be destroyed; it would seem that protons are not the immortal building blocks we once thought. Using the grand unified theories, we can estimate the lifetime, or average life expectancy, of the proton; it turns out to be $10^{32}$ years, which is much greater than the age of the Universe! This extremely long lifetime guarantees that although all matter is ultimately destined to disappear, the probability of decay in any given time span is exceedingly small. Nonetheless, given that our world is governed by statistical physics, any one proton is statistically in danger of decaying at any moment. In fact, since water is an abundant source of protons, theory predicts that on average one proton should decay per year in each ton of water. Experiments are now in progress attempting to detect such events in huge quantities of water stored in deep underground tanks at several places on Earth (thus shielding them from spurious effects triggered by cosmic rays reaching Earth's surface from outer space). Furthermore, a statistical measurement of a proton's lifetime should enable us to discriminate among the various GUTs, further refining our "approximations of reality." The simplest of the GUTs have apparently been ruled out as no proton decays have been found in the last decade in several tons of water. Some physicists take this as a bad sign, for the history of science has taught us that, more often than not, theoretical complications usually indicate that we are on the wrong track.

Another prediction of the inflationary scenario—and thus another indirect test of the many different GUTs—concerns the origins of galaxies, perhaps the greatest missing link in all of cosmic evolution. Here, theory suggests that any, even extremely small-scale fluctuations in the matter density, $\rho_m$, before inflation—an inevitable consequence of quantum physics—would be stretched or amplified by inflation to an extent now characterizing whole galaxies and clusters of galaxies. Thus the growth of gravitational instabilities, greatly aided by inflation, might have gradually led to the formation of self-gravitating collections of matter. Should this idea be correct, then the vast conglomerates of matter we see today as galaxies, galaxy clusters, and even the truly vast galaxy superclusters are the progeny of quantum fluctuations prevalent when the Universe was a mere $10^{-35}$ s old. The next generation of large telescopes capable of probing the most ancient material realms of the cosmos will soon be gathering data in an attempt to elucidate the origin of galaxies and thus testing the behavior of such primeval fluctuations. Somewhat ironically, with the physicists unable to build apparatae on Earth sufficiently energetic to reproduce cosmic chaos, it is the astronomers who, by studying the macrorealm, are beginning to provide tests, albeit indirect ones, of the grand unification of the microrealm.

An intriguing cosmological implication of the inflationary concept is that, if correct, it must have put the Universe into a state precariously balanced between infinite expansion (k = −1) and ultimate collapse (k = +1). Recall that for this to happen, $\rho_m = 3H^2/8\pi G$, which is the critical k = 0 case just sufficient for its accumulated gravitational effects to retard the rate of expansion. And since astronomers have observationally demonstrated that the density of normal matter (e.g., protons, neutrons, electrons) is only about 10% of this critical density, we surmise that upwards of 90% of the matter in the Universe is not ordinary, but in some unorthodox form such as black holes, massive neutrinos, or particles not yet known to physics.

What about even earlier phases of the chaos period—at t < $10^{-35}$ s? Can we probe, even theoretically, any closer to the celebrated t = 0 moment? Efforts are currently hampered because doing so requires the gravitational force to be incorporated into the correct GUT. Indeed, no one has yet succeeded in developing a super-grand unified theory (or "super-GUT"), as this is tantamount to inventing a quantum theory of gravity—an intellectual advance that requires a synthesis of Heisenberg's Uncertainty Principle and Einstein's Relativity Theory. Even so, our current knowledge of strong gravitational forces implies that such quantum effects very likely become important whenever the Universe is even more energetic than we have yet considered. Specifically, at t $\simeq 10^{-43}$ s (known as the "Planck time"), when T $\simeq 10^{32}$ K, the four known basic forces are thought to have been one—a truly fundamental force operating at energies characterizing the earliest parts of the chaos period. There and then, with all the matter in the Universe theorized to have been unimaginably compacted, both the curvature radius of (Einsteinian) spacetime and the distance (Heisenberg) uncertainty equal $10^{-33}$ cm (the "Planck length"), inside which relativity theory is no longer an adequate description of Nature. Only at smaller energies (i.e., at $t > 10^{-43}$ s) would the more familiar four forces begin to manifest themselves distinctly, though in reality all four are presumably different aspects of the single, fundamental, super-grand force that ruled at or near the big bang.

In a recent and potentially relevant development, some researchers have become enamored with a radical idea proposed some three decades ago. Called "superstrings," this theory aspires to unite all the laws of physics into a single mathematical framework. The name derives from the novel idea that the ultimate building blocks of Nature are not point particles at all, but tiny vibrating string-like entities. If this view is correct, it means that the protons and neutrons in all matter, from our bodies to the farthest star, are fundamentally made of strings. However, no one has ever seen such strings since they are predicted to be more than a billion billion times smaller than a proton—in fact, $10^{-33}$ cm, the Planck length. Depending on the mode of vibration, separate particles of matter can be created from the subatomic strings, much the way a violin string can resonate with different frequencies, each one creating a separate tune of the musical scale. Disconcertingly, the theory of superstrings works only if the Universe began with ten dimensions, six of which (somehow) become "hidden" near the time of the big bang. To some physicists, such a revolutionary idea borders on science fiction (and even theology), whereas for others it possesses breathtaking elegance. Even so, the world of science is littered with mathematically elegant theories that apparently have no basis in physical reality. And although the theory of superstrings is now causing great excitement in the physics community, there is to date not a shred of experimental or observational evidence to support it.

To penetrate even closer to the beginning of all time is currently hardly more than conjecture, though many researchers suspect that once a proper theory of quantum gravity is in hand, our understanding might automatically include a *natural* description of the original creation event itself. To this end, it is not inconceivable that the primal energy emerged at $t = 0$ from quite literally nothing. This might be true because even in a perfect or "zero-energy" vacuum—a region of space containing neither matter nor energy—particle-antiparticle pairs (such as an electron and its antiparticle opposite, the positron) are constantly created and annihilated in a time too short to observe. Although it would seem impossible that a particle could materialize from nothing, not even from energy, it so happens that no laws of physics are violated because the particle is annihilated by its corresponding antiparticle before either one can be detected. Furthermore, for such events not to happen would violate the laws of quantum physics, which cite, via the Heisenberg principle, the impossibility of determining exactly the energy content of a system at every moment in time. Hence, natural fluctuations in energy content must occur *even when the average energy present is zero.*

In this way, the Universe may well have originated from nothing by means of an energy change that lasted for an unimaginably short duration—a "self-creating Universe" that erupted into existence spontaneously, much as elementary particles occasionally and suddenly originate from nowhere during certain subnuclear reactions. Could this be the solution to Leibniz's philosophical query, "Why is there something instead of nothing?" The answer, ostensibly, is that the probability is greater that "something" rather than "nothing" will happen. Clearly the development of a quantum gravitational description of events at t = 0, which has thus far met with little success, is the foremost challenge in the subject of physics today. (Much of this long note derives from an earlier summary of the primordial Universe found in Field and Chaisson, 1985.)

4. Originally, in the Greek sense of the word, "chaos" referred to the formless, equilibrated entity from which the subsequently ordered Universe arose. Here, we take chaos to mean, in the more modern sense of the word, unordered confusion—the disorder or incoherence implied by its ordinary, everyday usage.

The term "chaos" as used in this book should be distinguished from the avant-garde mathematical subject of chaos theory that addresses the erratic, stochastic behavior of dynamical irregularities in Nature. Chaos theory itself has become an ambiguous, inconclusive description of what is also, and more properly, called dynamical systems theory. No universal agreement yet exists on what chaos theory is, or how it is that unpredictable behavior can reveal underlying order. Not to be blinded by the fashionable buzzword, chaos, which in its modern mathematical context actually embodies exquisite organization, this book addresses the search for order emerging from complex, non-linear dynamical systems, including those sometimes described by the paradoxical phrase, "deterministic chaos." In short, our study herein incorporates the essence of chaos theory as now vaguely understood, even though we have not chosen to call it that, or even to use that term anywhere in the text in its current mathematical sense. (See Ruelle 1991 in Further Reading.)

5. Neutrinos (ν), whether massless or slightly massive, decoupled from all other forms of matter and from radiation when T $\simeq 10^{10}$ K, the average temperature of the prevailing conditions at t $\simeq$ 1 s. Essentially, the density of matter, $\rho_m \simeq 10^{10}$ g cm$^{-3}$, had decreased enough to allow the neutrinos to roam freely; owing to their extremely weak interaction with other particles (including an ability to penetrate several light-years of lead without typically hitting any atoms), neutrinos were able to escape somewhat earlier than photons from the bulk matter comprising the early Universe. The primordial neutrinos (as well as their anti-neutrino counterparts) are reasoned to exist around us today as part of a "neutrino background" in the same numbers (~400 particles cm$^{-3}$) and at nearly the same temperature ($T_\nu = [4/11]^{1/3}$ $T_\gamma \simeq$ 2 K) as the primordial photons comprising the cosmic background radiation.

The issue of neutrino mass is as elusive as the particles themselves. (During the course of writing this book, a half-dozen claims and counterclaims were announced by the particle-physics community. The most recent, in 1998, was labeled "a compelling, definitive proof" that neutrinos do have mass, although that mass is apparently so small as to be undetermined still.) If neutrinos do have some mass, even a minute amount on the order of an equivalent mass of 5 electron volts (eV) each, or ~$10^{-11}$ erg, which is ~100,000 times less than that of an electron (the lightest known particle having any mass of its own), then given the neutrino's great numbers (comparable to the photon-to-baryon number density ratio $\sim 10^9$), they would comprise a considerable fraction of the dark matter in the Universe. This would allow neutrinos to make a moderate contribution to the energy density *of matter,* $\rho_m c^2$; if $H_0 \simeq 65$ km s$^{-1}$ Mpc$^{-1}$, neutrinos would, however, need a mass of nearly 50 eV ($\sim 10^{-10}$ erg) to close the Universe. Such mass would also grant them a significant role in the gravitational assembly of the largest known objects, especially the origin and evolution of galaxies and galaxy clusters on grand scales. Yet, such mass would alter but a few details of the historical narrative presented here, and not much at all of the big-picture scenario of cosmic evolution. If, by contrast, the neutrinos have no mass at all (like the photons), these ghostlike particles are destined to have no effect on large-scale structure, to contribute nothing to energy density, and to interact with matter only sparingly and in special places, such as the dense cores of massive stars.

6. Some researchers call this initial cosmic period the Energy Era. Indeed I did, too, in an earlier book for general readers (Chaisson 1989b), although in a textbook (Chaisson 1988; both listed in Further Reading), it was termed the Radiation Era. Given that matter is energy, and that life which is matter is also energy, it is more appropriate to label this earliest time frame the Radiation Era, since radiative photons (the purest form of energy) then dominated. Accordingly, the concept of energy (including its associated terms such as energy density) is the common denominator linking radiation, matter, and life throughout each of the major eras in the history of the Universe.

7. Hydrogen was not the only kind of atom formed in the early Universe. Since small quantities of slightly heavier nuclei had been synthesized in the prior nuclear epoch, they too were able to attract the appropriate numbers of electrons to form neutral atoms. Accordingly, all cosmic objects should contain at least 8% helium abundance by number, or at least 24% by mass. For two reasons, this is an irreducible minimum amount of helium that should contaminate virtually all objects in the Universe. First, physical processes such as the proton-proton cycle within stars have surely created additional amounts of helium well after the big bang. And second, since helium is chemically inert, it cannot be easily changed into something else once formed; helium atoms cannot even "hide" within other substances, like molecules, since helium does not easily combine with other elements.

Small rocky planets are exceptions. They usually have little helium because their gravity isn't strong enough to prevent helium atoms from escaping. Life is also an exception, for it likely originated from the matter of which our planet is made. At any rate, the fact that the oldest stars, especially those within globular clusters, contain just about 10% of their atoms in the form of helium lends support to the idea that most of the helium was indeed created in the early moments of an explosive Universe. In fact, the minimum and especially uniform abundance of $^4$He observed just about everywhere examined in the Universe is considered strong corroborating evidence for the primordial big-bang concept, since the process of nucleosynthesis within stars (which otherwise works fine for the production of heavy elements) is hard pressed to explain both the large abundance and uniformity of this lightweight element. Note also that because of helium's ionization potentials (54 eV for double ionization, 24 eV for single ionization), neutral helium atoms would have actually formed prior to neutral hydrogen atoms (whose ionization potential is 13.6 eV), when $T \simeq 200,000$ K.

8. The grand event of universal decoupling was somewhat gradual, if only because of the structure of hydrogen atoms. Some 13.6 eV are needed to ionize hydrogen completely; $2.7 \, k_B T = 13.6$ eV when $T \simeq 10^5$ K, and this occurred in the early Universe at $t \simeq 10^4$ years, just about when the Radiation Era subsided. At this temperature, although H atoms were beginning to form, they were still excited atoms; and such bound though excited states can still scatter photons (much as free, unbound electrons had done very efficiently in the earlier Universe). Only when the captured electron has reached the ground state, do the H atoms cease to interact significantly with radiation. Such minimum-energy H atoms had to await further cooling, to about 4000 K at $t \simeq 10^5$ years, for the matter and radiation to decouple fully, and for the Universe to become truly transparent. By this, we mean that the "mean free path" of a photon—the average distance a photon travels between scatterings by two charged particles—became much longer than the distance traversed by photons during the characteristic expansion time, $H_0^{-1}$, of the Universe for that period. An apt analogy is the way we view the Sun today: Within the solar interior, the gas is ionized and excited, causing so much scattering of the photons directly that it takes about a million years for them to random-walk their way to the surface photosphere. The photons we do see are in the form of sunlight that arises from that photosphere, from which those photons stream directly toward the Earth, reaching here in 8 minutes; the solar photosphere acts as a "surface of last scattering," after which radiation travels through the effectively "transparent" interplanetary space to Earth.

Researchers often label this decoupling event "recombination," as it is labeled in this book. This is a perfectly good term to use to describe later events, such as those that occur in gaseous nebulae where the radiation of bright young stars ionize surrounding atoms, after which some atoms do in fact "recombine." But in the intensely hot early Universe, there were no atoms prior to the formation of the first ones, thus the term "combination" is more semantically appropriate.

9. Theorists reason that before the change from the Radiation Era to the Matter Era, two distinct types of fluctuations could have been present in the gas of the early Universe. The first type of fluctuation maintains a constant temperature throughout the eddy; these are technically termed "isothermal" fluctuations since no temperature change is involved. The second, "adiabatic" fluctuations are disturbances that vary in step with the density of the gas but have no heat transferred within the eddy. The behavior and evolution of the two types of disturbances differ greatly.

On the one hand, adiabatic disturbances within eddies containing $<10^{12}$ M$_\odot$ would have dissipated with time; only those denser-than-average eddies housing $>10^{12}$ M$_\odot$ could have survived the great change from the Radiation to the Matter Era. The reason for this lies in the behavior of the radiation trapped in the slightly denser eddies. Because it exerts pressure, the radiation causes an oscillation in the gas density, just as in a sound wave. But the trapping is not complete if the eddy is too small, and the radiation diffuses away after a few oscillations—and with it the eddy. Only if the eddy is large enough can this type of oscillation survive until the Matter Era (and especially the galaxy epoch) begin. This is a top-down approach, suggesting that individual galaxies fragmented from much larger clouds that eventually gave birth to whole clusters of galaxies.

On the other hand, isothermal disturbances containing relatively small amounts of matter could have survived throughout the Radiation Era; the pressure of radiation in such fluctuations is constant throughout space anyway, and the matter is distributed independently. According to this bottom-up approach, moderate-sized clouds ($10^{6-8}$ $M_\odot$, smaller than normal galaxies) form first, after which they merge to become galaxies; as time passes, the galaxies themselves agglomerate to form larger units, such as galaxy clusters, sheets, and filaments. Because computer-modeling results appear to mimic the real Universe, many astronomers have embraced the notion that galaxies really did form from isothermal fluctuations in the early Universe.

But closer inspection of the distribution of galaxies in space suggests a problem. Observations of the vast clusters of galaxies clearly show a tendency for elliptical galaxies to be found in regions where the numbers of galaxies are greatest, whereas spirals are usually found where the numbers of galaxies are low. If individual galaxies formed first and the galaxy clusters formed later, why would the ellipticals be segregated from the spirals? After all, the clustering ability of two or more galaxies depends only on gravity, and this long-range force does not care whether the galaxies are spiral or elliptical. These recent observations, along with current conjecture about the very early Universe, imply that individual galaxies more likely formed only after the huge clouds of which they were a part were already quite compressed—the huge clouds being the parents of the immense galaxy clusters.

No one currently knows which, if either, scenario of galaxy formation is correct. Recent observations of small-scale distortions in the cosmic background radiation, almost surely present before recombination (at $t \simeq 100,000$ years), show much promise regarding the study of the origins of galaxies. Despite considerable efforts, however, the specifics of a plausible galaxy formation process have thus far eluded discovery.

## 3. LIFE

1. The definition of life given in the text (and Glossary) is my preferred version, intended to emphasize similarities rather than differences between animate and inanimate structures. Many other definitions of life have been offered by other researchers; without attempting any formal attributions, here are some notable ones collected over the years:

- Life is a biological organism that can sustain its own existence.

- Life is something that shares the characteristics of metabolism, reproduction, and growth.

- Life is a system capable of exercising function, such as breathing, movement, and reproduction.

- Life is a system made of organisms that can reproduce and transmit their genes to successive generations.

- Life is a bounded, informed, self-replicating, autocatalytic, dissipative structure.

- Life is a self-sustained, chemical system capable of undergoing Darwinian evolution.

- Life is a group of chemical systems in which free energy, released as part of the reactions of one or more of the systems, is used in the reactions of one or more of the remaining systems.

- Life is a potentially self-perpetuating open system of linked organic reactions, catalyzed stepwise and almost isothermally by complex and specific organic catalysts, which are themselves produced by the system.

Contemporary biology has no simple, clean definition of life agreed upon by all practitioners. In contrast to a definition of matter, for which a dozen physicists from different countries would give basically (and perhaps precisely) the same answer—"anything that occupies space and has mass"—a dozen biologists (in the same room, let alone from different countries) would likely produce nearly as many different definitions of life. This is all the more reason to consider seriously the long, yet unreserved, thermodynamically oriented definition of life given in the text. Our interdisciplinary agenda is to soften the boundaries between physics and biology, indeed among all the natural sciences.

2. Even if loose neutral matter did reionize (at $z \simeq 50$), the matter and radiation temperatures would never again be equalized (unless the Universe does someday collapse), ensuring the disequilibrium conditions presumably needed for the onset of order, flow, and complexity in the Universe. This intergalactic matter must have had an integral role regarding the origin of the galaxies early on, and even now might persist in the form of significant amounts of dark matter. In fact, however, only small amounts of true (beyond-galaxy-clusters) intergalactic matter have been observationally found to date, either directly as diffuse gas radiating in the 21-cm radio band or indirectly as foreground-absorbing clouds (the "Lyman-$\alpha$ forest") in the optical spectra of highly red-shifted quasars. The status, alas even the existence, of true intergalactic matter, now and throughout the history of the Universe, remains unclear. At any rate, for the purposes of this work, its thermal behavior differs from that of photons, once decoupled (at $z \simeq 1500$); except for the first few thousand centuries, matter and radiation have differed enough in average temperature for us to be able to consider the entire Universe as a vast heat engine. We shall proceed accordingly in our analysis, showing the effect of such a thermal divergence between matter and radiation, in principle irrespective of whether $T_m > T_r$, or conversely. It is the *existence* of a thermal gradient that counts most, whatever its cause.

3. A value of $\Phi_m$ of order unity holds true for all stars, not just for our Sun, pro-

vided they are normal stars on the main sequence (see Figure 21). For example, the giant blue star Vega has some fifty times larger luminosity, yet has many times more mass, making its $\Phi_m$ value similar to that of the Sun. Even supergiant main-sequence stars having thousands of solar luminosities also have masses of the order of a hundred solar masses, making $\Phi_m$ within a factor of ten of unity. Likewise, red-dwarf stars such as ε Eridani have 0.3 $L_\odot$ and ~0.5 $M_\odot$, once more making $\Phi_m$ comparable to the Sun's value; the extremely dim subdwarf stars do have $\Phi_m$ values much less, probably owing to their near-equilibrium states displaying weak thermal and chemical gradients.

In the Discussion, we shall return to note the larger variations in $\Phi_m$ among extreme-state stars that have left the main sequence of normal stars. The larger range in $\Phi_m$ for evolved stars will actually be a measure of their stellar evolution and of the increased complexity among those evolved stars.

4. $\Phi_m = 0.1–1 \ L_\odot M_\odot^{-1}$ holds for most normal (spiral and elliptical) galaxies, although some active galaxies (such as the enigmatic quasars) display values of $\Phi_m$ upwards of 100 times larger, at least for short periods of time and in selected regions—much as fanatical cyclists at the Tour de France or race horses in the Kentucky Derby can reach metabolic rates 6 and 20 times their respective basal (resting) metabolic rates. Values of $\Phi_m$ can therefore have considerable variation within a given type of organized system, depending upon the degree of activity, requiring care when comparing values of $\Phi_m$ among different systems.

We also elect not to address, at this time, systems larger than typical galaxies. While observed L/M ratios for normal galaxies average 0.5 erg s$^{-1}$ g$^{-1}$, those ratios decrease by one and two orders of magnitude for galaxy clusters and superclusters, respectively, implying significant amounts of gravitating material over and above what can be observed through the radiation it produces. In short, the larger the gravitationally bound system, the greater the dark-matter problem. Until the dark-matter issue is resolved, or at least better understood, the kind of analysis suggested in this book is best postponed.

5. The current value of $\Phi_m$ for the entire Earth, as a rocky planet third out from the Sun, is largely irrelevant to our present argument, for the Earth in bulk is not now evolving appreciably. Earth's energy flow, mostly in the form of radiogenic heat upwelling from within—left over from its gravitational contraction and accretion billions of years ago, from its heavy meteoritic bombardment (as inferred from lunar cratering and dating), and from its subsequent radioactive decay—is currently small. Its total heat outflow at the surface is measured and globally averaged to be ~63 erg cm$^{-2}$ s$^{-1}$ (Hubbard 1984), which when integrated over the entire surface of the globe translates into an effective (geothermal) luminosity of ~3 × 10$^{20}$ erg s$^{-1}$. That makes $\Phi_m$ for Earth en masse currently ~5 × 10$^{-8}$ erg s$^{-1}$ g$^{-1}$, a value so small as to be consistent with an ordered yet virtually unchanging physical object, like an already formed, yet dormant, crystal having $\Phi_m \simeq 0$—which, by the way, much of Earth is. Even this small heat flow, however, can affect planetary evolution at the surface and drive events with implications for life; tectonic activity represented by recent mountain-building or volcanism such as the Alps or the western United States has a current value of $\Phi_m$ twice that of geologically old and inac-

tive areas such as Pre-Cambrian shields. As perhaps expected, mid-oceanic trenches are the sites of greatest radiogenic heat flow at or near the surface of Earth today, reaching values of $\sim$150 erg cm$^{-2}$ s$^{-1}$, and sometimes double that in especially energetic vents.

Earlier in Earth's history, when it was changing more rapidly during its first billion years or so—developing, settling, heating up, differentiating—its value of $\Phi_m$ would have been much larger. Taking a surface temperature $T_s \simeq 1800$ K (Hartmann 1993) as an average value of a "magma ocean" during its initial half-billion years, and knowing that energy flux scales as $\sigma T^4$ (where $\sigma$ is the Stefan-Boltzmann constant), we can estimate that in Earth's formative years, its free energy rate density would have been enhanced by $(1800/255)^4$, making $\Phi_m$ *then* several orders of magnitude larger than now. (We take 255 K, not 290 K; the former is the "balanced temperature" of the planet, where the incoming solar energy absorbed equals the outgoing terrestrial heat emitted, which would have been more indicative of early Earth, whereas the latter is the "greenhouse temperature" boosted in more recent times by the thickening of Earth's atmosphere.)

Even earlier, when primordial Earth accumulated a vast amount of heat from the energy of accretion (the conversion of gravitational potential energy into heat during the act of formation), its value of $\Phi_m$ would have been of order tenish erg s$^{-1}$ g$^{-1}$, as we later show in the Discussion. This ancient value of $\Phi_m$ accords well with our energy flow diagnostics, namely is larger than that of a less ordered star yet smaller than that of a more-ordered climasphere. This, then, would have been the flow of energy through Earth proper when it was experiencing its most dramatic ordering phase, virtually completed within a fraction of its first billion years. Much of the early planet's structure would have been driven by ancient $\Phi_m$ and then rather quickly set in its patterned ways of layered stratification, core rotation, and mantle convection, after which little further ordering has occurred except at the surface—where, not coincidentally, the cosmic evolutionary story continues.

6. As with stars and planets, the range of $\Phi_m$ values for living systems can vary considerably, sometimes over an order of magnitude or more. Photosynthesis is not normally an efficient process, converting only $\sim$0.1% of the incoming solar energy into chemical energy stored in glucose molecules; its low efficiency is actually due more to limiting supplies of atmospheric $CO_2$ than to a lack of energy. Our value for $\Phi_m$ used in Chapter Three, valid for the bulk of Earth's flora, namely uncultivated crops, will be reconsidered in the Discussion among a range of $\Phi_m$ for more organized, evolved plants.

7. We use the thermochemical definition of a calorie, 1 cal $= 4.184 \times 10^7$ erg, not the dietician's value (also known as a large Calorie), which is a thousand times more energetic. Also, the energy budget is here assumed for today's average sedentary human, which is some 60% higher than the basal metabolic rate of 1640 kcal day$^{-1}$ for a 70-kg person fasting and lying motionless in bed all day and night. By contrast, and much akin to the active galaxies that, for relatively brief durations, display values of $\Phi_m$ equal to nearly a hundred times the normal value for galaxies, humans can require several tens of times their basal rate during short periods of

maximum exertion, or high-endurance athletics, when power expenditures can reach a few thousand watts.

8. As with other structures, and not just organisms, human brains have $\Phi_m$ values that vary somewhat depending on level of development. In the text, we have considered a mature adult brain, which typically consumes ~20% of the body's total energy intake. By contrast, the brain of a newborn child consumes ~60% of the energy acquired, a not surprising result given that a body's lump of gray matter doubles in size during the first year of life and synapse formation is most dramatic in the pre-school years. Thus, $\Phi_m$ can average several times larger for an infant brain than for an adult brain. This pattern we have seen before: For planets, a good deal more energy was needed during formative development, but is needed less now to maintain them (see note 5, this chapter). And we shall encounter it again since, during ontological development, most organisms change from a higher to lower metabolic rate.

9. In calculating $\Phi_m$ for contemporary society, only a mass of humankind tout ensemble is used. We have not included the mass of modern civilization's infrastructure—buildings, roadways, vehicles, and so on—any more than we have included the mass of the human body when calculating $\Phi_m$ for the human brain, or that of the host Galaxy when considering the Sun. While attempting to assess the degree of a system's complexity, it is reasonable and proper to consider systems separate from their environments, just as we have consistently done in our earlier thermodynamic reasoning. For as with all systems studied in this book, the sum of order and complexity for a system *and* its environment will always be unimpressively low (in fact technically negative) in accord with the second law of thermodynamics.

10. Previous researchers, while studying flow rates in complex systems, have sought to identify maximum and minimum principles guiding the growth and evolution of those systems. For example, Lotka (1924) claimed that ecosystems evolve so as to maximize the total system throughput—a "principle of maximum power production"—whereas Prigogine (1980) argued that certain non-equilibrium systems have the lowest possible entropy generation for a given set of conditions—a "principle of minimum entropy production." By contrast, this work has embraced neither maximum nor minimum principles that have often been a hallmark of rigid reductionistic science. Instead, we have opted for optimization strategies suggestive of a more liberal (holistic?) interpretation of systems interacting with their environments—criteria that best suit organized systems employing moderate flows of energy large enough to sustain them yet small enough not to destroy them. No unambiguous extremum principles have been found in this study, nor are any likely to hold over such a wide range of ordered structures, from primordial atoms to sentient beings.

11. Despite our disavowal of a flame being alive, it is interesting that many strong comparisons have been made between fire and life. Leonardo da Vinci drew prominent analogies between flames and the living process; firefighters do it today, often swearing that flames are cunning, deceptive, and downright intelligent; many others see in a flame the reproductive process, for flames surely do multiply. Their simi-

larities go deeper, too: A flame, having a $\Phi_m$ value comparable to much of life, burns fuel, again like a biological cell, to yield energy, $CO_2$ and $H_2O$. However, there are some clear differences as well: A flame undergoes uncontrolled combustion, compared to the oxidation of a typical cell that experiences rigorous control. What is more, the flame does so in one step, transforming incoming chemical energy directly into heat; by contrast, a cell requires many steps but little loss of heat to convert chemical energy into many forms of retained energy, which then serves to drive molecular reactions as well as to build and maintain structure. In either case, flame or cell, the high-entropy product—the heat released—gets dumped into the environment, thereby increasing the entropy of the surroundings as demanded by the second law of thermodynamics.

## DISCUSSION

1. Different types of equilibria need to be distinguished. Despite the clear thermal and chemical disequilibrium extending from core to surface of any functioning star, such stars do remain hydrostatically equilibrated between gravity pulling in and pressure pushing out. In this way, they maintain just enough structural integrity to remain reasonably stable as stars, lest they otherwise blow up—at least until suffering cataclysmic "death" when many of them (unlike the Sun) do in fact explode as supernovae. Such an explosion is a star's way of returning, violently, to a genuine equilibrium state with its surrounding interstellar environment. The complexity of a detonated star itself is surely diminished, yet that of the galactic medium is just as surely increased by means of an enrichment of heavy elements.

We take this opportunity also to note that gravity is the trigger, nuclear reactions merely the result. It is gravitational infall that creates the high temperatures at the nascent star's core, hence the crucial temperature gradient needed to get the energy flowing. The nuclear reactions subsequently ignite because such stars convert gravitational potential energy to gas kinetic energy and then to vast luminosities that escape. That's also why the most massive stars fuse faster and, somewhat contrary to expectations, endure less; their greater masses push down harder, cause higher core temperatures, accelerate the nuclear reaction rates, and produce even greater energy flows.

2. Stars like the Sun increase their luminosity through time as hydrogen in their cores converts to helium. Measurements of the solar constant ($\propto$ energy reaching Earth) imply that the Sun is currently brightening by ~1% every 100 million years. Extrapolating back some 3–4 billion years, the early faint Sun was probably only a third as luminous as today—an estimate that agrees with recent geophysical evidence of early Earth conditions (see Sagan and Chyba 1997). Despite the Sun's early faintness, Earth might have been warmed by greenhouse heating caused by the release of even minute amounts of ammonia from undersea hydrothermal vents.

3. A typical daily agenda for a 70-kg human being can be broken down as follows, noting that $\Phi_m$, our complexity indicator, does indeed rise with task difficulty:

| Human Activity | Duration (hours) | Rate (calories/min) | $\Phi_m$ (erg s$^{-1}$ g$^{-1}$) | Total (calories) |
|---|---|---|---|---|
| sleeping/lying (basal) | 8 | 1.1 | 11,000 | 540 |
| sitting (reading, eating) | 6 | 1.5 | 15,000 | 540 |
| standing (at ease, indoors) | 6 | 2.4 | 23,800 | 860 |
| walking (steady, outdoors) | 2 | 2.8 | 27,800 | 320 |
| other (sports, exercise) | 2 | 4.5 | 44,700 | 540 |
| | | | Total = | 2,800 cal/day |

4. Jean-Baptiste de Lamarck, in his masterwork, *Philosophie zoologique* (1809), developed the idea of "inheritance of acquired characters" and is often considered the founder of modern evolutionism. Favoring the uniformitarian (or slow-change), as opposed to catastrophic (or violent-change), school of geology, Lamarck repeatedly stressed the gradual and even sluggish tenor of the events that yield change, both physically and biologically. Lamarck's central thesis has come to be known as the law of use and disuse, whereby the environment does not directly (and therefore rapidly) produce changes in a life form; rather, Nature engenders gradual change indirectly according to a life form's reaction toward its environment. In the classic example cited in most of today's biology textbooks, Lamarckism maintains that giraffes have long necks because the necks of some of their predecessors were used extensively, even stretched in order to reach the leaves high in the trees of the African plains, after which the long necks were inherited. Similarly, woolly mammoths produced more hair in cold climates because in glacial times parents grew thicker hair and transmitted hereditary potential for hair growth to their offspring. Little or no genetic selection is involved in Lamarckism as the dynamic regards the individual. However, controlled experiments have since strongly suggested that such contributions are factually incorrect; muscles made large by much use are not passed on to the next generation of offspring. Given neo-Darwinism's later success in accounting for life's many varied changes by means of natural selection within a population of individuals, Lamarck's explanation for the mechanism of biological evolution is no longer tenable within the scientific community; rather, we accept, with Darwin, that those giraffes within a larger population of giraffes that are genetically endowed with longer necks are those that survive better and produce more progeny because they had the advantage of reaching the higher leaves. By contrast, much of cultural evolution, which follows biological evolution and does incorporate the passage of traits and factors that made us human, obeys (at least in part) Lamarckism; the "good" traits invented and accumulated by one generation are inherited by the next generation by means of systems that favor schooling, memory, tradition, and other such social contracts that communicate knowledge.

In sum, Lamarckism puts the emphasis on isolated, individual organisms; in principle, this mechanism could work even if no more than a single individual were alive at any one time. By contrast, Darwinism, which is a function of populational change, cannot possibly work with only one organism; it is a group concept. In fact, perhaps the most distinguishing characteristic of Darwinism is that it works on populations, not individuals. This explains in part why Darwinism dominates biological evolution, Lamarckism cultural evolution.

# WORKS CITED

Adams, R. N., *Energy and Structure,* 1975, Univ. of Texas Press, Austin.

Angelopoulos, A., et al., "First Direct Observations of Time-Reversal Non-Invariance in the Neutral-Kaon System," *Physics Letters,* vB444, p. 43, 1998.

Ayres, R. U. *Information, Entropy, and Progress,* 1994, Amer. Inst. of Phys. Press, New York.

Bak, P., *How Nature Works,* 1996, Springer-Verlag, New York.

Barrow, J. D. and Tipler, F. J., *The Anthropic Cosmological Principle,* 1986, Oxford Univ. Press, Oxford.

Bennett, C. H., "The Thermodynamics of Computation—a Review," *Int. J. Theoretical Physics,* v. 21, p. 905, 1982.

Bent, H. A., *The Second Law,* 1965, Oxford Univ. Press, Oxford.

Bergson, H., *Evolution Créatrice,* [1907] 1940, Presses Universitaires de France, Paris.

Bertalanffy, L. von, *Theoretische Biologie,* v. 1, 1932, Borntraeger, Berlin.

Blum, H. F., *Time's Arrow and Evolution,* 1968, Princeton Univ. Press, Princeton.

Bonner, J. T., *The Evolution of Complexity,* 1988, Princeton Univ. Press, Princeton.

Brillouin, L., *Science and Information Theory,* 1962, Academic Press, New York.

Brooks, D. R. and Wiley, E. O., *Evolution as Entropy,* 1988, Univ. of Chicago Press, Chicago.

Brown, H., "Energy in Our Future," *Ann. Rev. Energy,* v. 1, p. 1, 1976.

Caplan, S. R. and Essig, A., *Bioenergetics and Linear Nonequilibrium Thermodynamics,* 1983, Harvard Univ. Press, Cambridge, MA.

Careri, G., *Order and Disorder in Matter,* 1984, Benjamin/Cummings, Menlo Park, California.

Carnot, S., *Reflections on the Motive Power of Fire,* 1824; rpt. 1986, Manchester Univ. Press, Manchester.

Carter, B., "The Anthropic Principle," in *Confrontation of Cosmological Theories with Observation,* M. S. Longair (ed.), 1974, Reidel, Berlin.

Cech, T. R., "Self-splicing of Group 1 Introns," *Ann. Rev. Biochemistry,* v. 59, p. 543, 1990.

Chaisson, E. J., "Our Cosmic Heritage," *Zygon,* v. 23, p. 469, 1988.

——, "Cosmic Age Controversy Is Overstated," *Science,* v. 276, p. 1089, 1997a.

——, "The Rise of Information in an Evolutionary Universe," *World Futures: The Journal of General Evolution,* v. 50, p. 447, 1997b; similar version in *The Quest for a Unified Theory of Information,* W. Hofkirckner (ed.), 1998, Gordon & Breach, Amsterdam.

——, "The Cosmic Environment for the Growth of Complexity," *BioSystems,* v. 46, p. 13, 1998.

——, "Ethical Evolution," *Zygon,* v. 34, p. 265, 1999.

Chaisson, E. and McMillan, S., *Astronomy Today* (3rd ed.), 2000, Prentice-Hall, Upper Saddle River, NJ.

Chandrasekhar, S., *Hydrodynamic and Hydromagnetic Stability,* 1961, Clarendon Press, Oxford.

Chothia, C., "One Thousand Families for the Molecular Biologist," *Nature,* v. 357, p. 543, 1992.

Coles, P. and Ellis, G. F. R., *Is the Universe Open or Closed?,* 1997, Cambridge Univ. Press, Cambridge.

Comte, A., *Positive Philosophy,* 1842, Geo. Bell, Montpellier.

Cook, E., "The Flow of Energy in an Industrial Society," *Scientific American,* v. 224, p. 135, 1971.

Corliss, J. B., Baross, J. A., Hoffman, S. E., "An Hypothesis Concerning the Relationship between Submarine Hot Springs and the Origin of Life on Earth," *Oceanologica Acta,* v. 4, p. 59, 1981.

Corning, P. A., "Complexity is Just a Word!," *Technological Forecasting and Social Change,* v. 59, p. 198, 1998.

Corning, P. A. and Kline, S. J., "Thermodynamics, Information and Life Revisited, Part I: 'To Be or Entropy'", *Systems Research,* v. 15, p. 273, 1998a.

——, "Thermodynamics, Information and Life Revisited, Part II: 'Thermo-economics and Control Information,'" *Systems Research,* v. 15, p. 453, 1998b.

Csányi, V., *Evolutionary Systems and Society,* 1989, Duke Univ. Press, Durham, NC.

Darwin, C., *On the Origin of Species,* 1859, J. Murray, London; facsimile of the first edition, 1964, Harvard Univ. Press, Cambridge, MA.

Davies, P. C. W., *The Cosmic Blueprint,* 1988, Simon & Schuster, New York.

Dawkins, R., *The Selfish Gene,* 1976, Oxford Univ. Press, Oxford.

——, *Climbing Mount Improbable,* 1996, Viking, London.

Day, R. H., "The General Theory of Disequilibrium in Economics and of Economics Evolution," *Economic Evolution and Structural Adjustment,* v. 293, p. 46, 1987.

Dennett, D. C., *Darwin's Dangerous Idea,* 1995, Simon & Schuster, New York.

Dobzhansky, T., Ayala, F. J., Stebbins, G. L., and Valentine, J. W., *Evolution,* 1977, W.H. Freeman, San Francisco.

Dollo, L., "Les Lois de l'Evolution," *Bull. Belge. Geol.*, v. 7, p. 164, 1893.

Dyke, C., "Cities as Dissipative Structures," in *Entropy, Information, and Evolution*, B. H.Weber et al. (eds.), 1988, MIT Press, Cambridge, MA.

Dyson, F. J., "Energy in the Universe," *Scientific American*, v. 225, p. 51, 1971.

———, "Time Without End: Physics and Biology in an Open Universe," *Reviews of Modern Physics*, v. 51, p. 447, 1979.

———, *Origins of Life*, 1985, Cambridge Univ. Press, Cambridge.

Eddington, A. S., *The Nature of the Physical World*, 1928, Cambridge Univ. Press, Cambridge.

Eigen, M., *Steps towards Life*, 1992, Oxford Univ. Press, Oxford.

Einstein, A., *The Meaning of Relativity*, 1922, Princeton Univ. Press, Princeton.

Eldredge, N., *Time Frames*, 1985, Simon and Schuster, New York.

Eldredge, N. and Gould, S. J., "Punctuated Equilibria: An alternative to phyletic gradualism," in *Models in Paleobiology*, T. J. M. Schopf (ed.), p. 82, 1972, W.H. Freeman, San Francisco.

Elmegreen, B. G. and Lada, C. J., "Sequential Formation of Subgroups in OB Associations," *Astrophys. J.*, v. 214, p. 725, 1977.

Emanuel, K. A., "Thermodynamic Control of Hurricane Intensity," *Nature*, v. 401, p. 665, 1999.

Eriksson, K.-E., Islam, S., and Skagerstam, B.-S., "A Model for the Cosmic Creation of Nuclear Energy," *Nature*, v. 296, p. 540, 1982.

Feynman, R., *The Character of Physical Law*, 1967, MIT Press, Cambridge, MA.

Feynman, R. P., Leighton, R. B., and Sands, M., *The Feynman Lectures in Physics*, v. 2, 1964, Addison-Wesley, Reading, MA.

Field, G. B. and Chaisson, E. J., *The Invisible Universe*, 1985, Birkhauser Boston; 1987 Vintage Books, Random House, New York.

Fortey, R., *Life: An Unauthorized Biography*, 1997, HarperCollins, London.

Fox, R. F., *Energy and the Evolution of Life*, 1988, W.H. Freeman, San Francisco.

Fox, S., *The Emergence of Life*, 1988, Basic Books, New York.

Fox, S. and Dose, K., *Molecular Evolution and the Origin of Life*, 1977, Marcel Dekker, New York.

Frautschi, S., "Entropy in an Expanding Universe," *Science*, v. 217, p. 593, 1982.

Gell-Mann, M., *The Quark and the Jaguar*, 1994, W.H. Freeman, New York.

Georgescu-Roegen, N., *The Entropy Law and the Economic Process*, 1971, Harvard Univ. Press, Cambridge, MA.

Gesteland, R. F., Cech, T. R. and Atkins, J. F. (eds.), *The RNA World*, 1999, Cold Spring Harbor Laboratory Press, Cold Spring Harbor, NY.

Gold, T., "The Arrow of Time," *American Journal of Physics*, v. 30, p. 403, 1962; see also *Recent Developments in General Relativity*, 1962, p. 225, Pergamon-Macmillan, New York.

Gould, S. J., "Is a New and General Theory of Evolution Emerging?" *Paleobiology*, v. 6, p. 119, 1980.

———, "Darwinism and the Expansion of Evolutionary Theory," *Science*, v. 216, p. 380, 1982.

———, *Full House*, 1996, Harmony, New York.

Grünbaum, A., "The Anisotropy of Time," in *The Nature of Time,* T. Gold (ed.), 1967, Cornell Univ. Press, Ithaca.

Guth, A., *The Inflationary Universe,* 1997, Perseus Press, New York.

Haiman, Z. and Loeb, A., "Observational Signature of the First Quasars," *Astrophys. J.,* v. 503, p. 505, 1998.

Haken, H., "Cooperative phenomena in systems far from thermal equilibrium and in nonphysical systems," *Reviews of Modern Physics,* v. 47, p. 67, 1975.

Halacy, D. S., *Earth, Water, Wind and Sun,* 1977, Harper and Row, New York.

Hammond, K. A. and Diamond, J., "Maximal Sustained Energy Budgets in Humans and Animals," *Nature,* v. 386, p. 457, 1997.

Hartmann, W. K., *Moons and Planets,* 1993, Wordsworth, CA.

Hawking, S., "Black Hole Explosions?", *Nature,* v. 248, p. 30, 1974.

Hodgman, C. D., et al., *Handbook of Chemistry and Physics,* any annual edition, Chemical Rubber Publishing Co., Cleveland.

Holland, H. D., *The Chemical Evolution of the Atmosphere and the Ocean,* 1984, Princeton Univ. Press, Princeton.

Hubbard, W. B., *Planetary Interiors,* 1984, van Nostrand, New York.

Jantsch, E., *The Self-Organizing Universe,* 1980, Pergamon, Oxford.

Jaynes, E. T., "Information Theory and Statistical Mechanics," *Physical Review,* v. 106, p. 620, 1957.

Jervis, R., *System Effects: Complexity in Political and Social Life,* 1997, Princeton Univ. Press, Princeton.

Johnson, L., "Thermodynamic Origin of Ecosystems," in *Entropy, Information, and Evolution,* B. H. Weber et al. (eds.), 1988, MIT Press, Cambridge, MA.

Kaler, J., *Stars,* Scientific American Library Books, 1998, W.H. Freeman, New York.

Kauffman, S., "Antichaos and Adaptation," *Scientific American,* v. 265, p. 78, 1991.

———, *The Origins of Order,* 1993, Oxford Univ. Press, Oxford.

Kimura, M., *The Neutral Theory of Molecular Evolution,* 1983, Cambridge Univ. Press, Cambridge.

Kleiber, M., *The Fire of Life,* 1961, Wiley, New York.

Lamarck, Jean-Baptiste, *Philosophie zoologique,* 1809, Paris.

Landauer, R., "Irreversibility and Heat Generation in the Computing Process," *IBM J. Research & Development,* v. 3, p. 183, 1961.

Landsberg, P. T., "Can Entropy and 'Order' Increase Together?" *Physics Letters,* v. 102A, p. 171, 1984.

Lavenda, B. H., *Thermodynamics of Extremes,* 1995, Albion, Chichester.

Layzer, D., "The Arrow of Time," *Scientific American,* v. 233, p. 56, 1975; see also *Astrophys. J.,* v. 206, p. 559, 1976.

———, "Growth of Order in the Universe," in *Entropy, Information, and Evolution,* B. H. Weber et al., (eds.), 1988, MIT Press, Cambridge, MA.

Lehninger, A. L., *Bioenergetics* (2nd ed.), 1971, Benjamin, Menlo Park.

———, *Biochemistry,* 1975, Worth, New York.

Lewin, R., *Complexity,* 1992, Macmillan, New York.

Lewis, G. N., "The Symmetry of Time in Physics," *Science,* v. 71, p. 569, 1930.

Lewis, J. S. and Prinn, R. G., *Planets and their Atmospheres,* 1984, Academic Press, Orlando.

Lorenz, K., *Behind the Mirror,* 1977, Harcourt, Brace, Jovanovich, New York.

Lotka, A. J., "Contribution to the Energetics of Evolution," *Proceedings National Academy Sciences,* v. 8, p. 147, 1922.

———, *Elements of Physical Biology,* 1924, Williams & Wilkins, Baltimore; republished as *Elements of Mathematical Biology,* 1956, Dover, New York.

Lovejoy, C. O., "Evolution of Man and its Implication for General Principles of the Evolution of Intelligence," in *Life in the Universe,* Billingham, J. (ed.), 1981, MIT Press, Cambridge, MA.

Lovelock, J., *Gaia,* 1979, Oxford Univ. Press, Oxford.

Luchinsky, D. G. and McClintock, P. V. E., "Irreversibility of Classical Fluctuations Studied in Analogue Electrical Circuits," *Nature,* v. 389, p. 463, 1997.

Margulis, L., *Origin of Eucaryotic Cells,* 1970, Yale Univ. Press, New Haven.

Margulis, L. and Sagan, D., *Microcosmos,* 1986, Simon and Schuster, New York.

Marijuan, P. C., Conrad, M., et al. (eds.), "First Conference on Foundations of Information Sciences: From Computers and Quantum Physics to Cells, Nervous Systems, and Societies," *BioSystems,* v. 38, pp. 87–257, 1996.

Matsuno, K., *Protobiology: Physical Basis of Biology,* CRC Press, 1989, Boca Raton, Florida.

Maturana, H. R. and Varela, F. J., *The Tree of Knowledge,* 1988, Shambhala, Boston.

Mayr, E., *The Growth of Biological Thought,* 1982, Harvard Univ. Press, Cambridge, MA.

———, *This is Biology,* 1997, Harvard Univ. Press, Cambridge, MA.

McMahon, T. A. and Bonner, J. T., *On Size and Life,* 1983, W.H. Freeman, San Francisco.

Miller, S. L., "A Production of Amino Acids under Possible Primitive Earth Conditions," *Science,* v. 117, p. 528, 1953.

Morowitz, H. J., *Energy Flow in Biology,* 1968, Academic Press, New York.

Morris, S. C., *The Crucible of Creation,* 1998, Oxford Univ. Press, Oxford.

Morrison, P., "A Thermodynamic Characterization of Self-Reproduction," *Reviews of Modern Physics,* v. 36, p. 517, 1964.

———, "Thermodynamics and the Origin of Life," in *Molecules in the Galactic Environment,* M. A. Gordon and L. E. Snyder (eds.), 1973, Wiley, New York.

Munitz, M. K. (ed.), *Theories of the Universe,* 1957, Free Press, New York.

Odum, H. T., *Systems Ecology,* 1983, Wiley Interscience, New York.

———, "Self-Organization, Transformity, and Information," *Science,* v. 242, p. 1132, 1988.

Poincaré, H., *Science et Méthòde,* 1914, Flammarion, Paris.

Penrose, R., *The Emperor's New Mind,* 1989, Oxford University Press, Oxford.

Prigogine, I., *From Being to Becoming,* 1980, W.H. Freeman, San Francisco.

———, as quoted in "Scientists See a Loophole in the Fatal Law of Physics," by M. W. Browne, *New York Times,* 29 May 1979.

Prigogine, I., Nicolis, G. and Babloyantz, A., "Thermodynamics of Evolution," *Physics Today*, v. 11, p. 23 and v. 12, p. 38, 1972.

Rees, M. J., "Origin of the Universe," in *Origins*, A. C. Fabian (ed.),1988, Cambridge University Press, Cambridge.

Reeves, H., *The Hour of Our Delight*, 1991, W.H. Freeman, New York.

Sabelli, H. C., Carlson-Sabelli, L., Patel, M., and Sugerman, A., "Dynamics and Psychodynamics: Process Foundations of Psychology," *J. of Mind and Behavior*, v. 18, p. 305, 1997.

Sagan, C., "Organic Chemistry and Biology of the Interstellar Medium," in *Molecules in the Galactic Environment*, M. A. Gordon and L. E. Snyder (eds.), 1973, Wiley, New York.

Sagan, C. and Chyba, C., "The Faint Sun Paradox," *Science*, v. 276, p. 1217, 1997.

Salk, J., "An Evolutionary Approach to World Problems," *UNESCO*, Paris, 1982.

Saslaw, W. C. and Hamilton, A. J. S., "Thermodynamics and Galaxy Clustering: Non-linear Theory of High Order Correlations," *Astrophysical J.*, v. 276, p. 13, 1984.

Schneider, E. D. and Kay, J. J., "Order from Disorder: The Thermodynamics of Complexity in Biology," in *What is Life: The Next Fifty Years*, M. Murphy and L. O'Neill (eds.), 1995, Cambridge Univ. Press, Cambridge.

Schopf, J. W., *Cradle of Life*, 1999, Princeton Univ. Press, Princeton.

Schrödinger, E., *What is Life?*, 1944, Cambridge University Press, Cambridge.

Shannon, C. E. and Weaver, W., *The Mathematical Theory of Communication*, 1949 (reprinted 1963), University of Illinois Press, Champaign-Urbana.

Shapiro, R., *Origins*, 1986, Summit Books, New York.

Shapley, H., *Flights from Chaos*, 1930, McGraw Hill, New York.

———, *Beyond the Observatory*, 1967, Scribner's, New York.

Shock, E. L., "Geochemical Constraints on the Origin of Organic Compounds in Hydrothermal Systems," *Orig. Life Evol. Biosphere*, v. 20, p. 331, 1990.

Simpson, G. G., *Tempo and Mode in Evolution*, 1944, Columbia Univ. Press, New York.

Smil, V., *Energies*, 1999, MIT Press, Cambridge, MA.

Spencer, H., *A System of Synthetic Philosophy*, 1862–96, v. 1, *First Principles*, 1862, Williams & Norgate, London..

Stryer, L., *Biochemistry* (3rd ed.), 1988, W.H. Freeman, San Francisco.

Szathmary, E. and Maynard Smith, J., "The Major Evolutionary Transitions," *Nature*, v. 374, p. 227, 1995.

Szilard, L., "Uber die Entropieverminderung in Einem Thermodynamischen System bei Eingriffen Intelligenter Wesen," *Zeitschrift fur Physik*, v. 53, p. 840, 1929.

Thorne, K. S., *Black Holes and Time Warps*, 1994, W.W. Norton, New York.

Turing, A. M., "The Chemical Basis of Morphogenesis," *Philosophical Transaction Royal Society London*, Ser. B, v. 237, p. 37, 1952.

Wald, G., "The Origin of Life," *Scientific American*, v. 191, p. 44, 1954.

———, "Life and Mind in the Universe," *Int. J. of Quantum Chemistry*, v. 11, p. 1, 1984.

Weinberg, S., *Dreams of a Final Theory,* 1994, Vintage, New York.

West, G. B., Brown, J. H. and Enquist, B. J., "The Fourth Dimension of Life: Fractal Geometry and Allometric Scaling of Organisms," *Science,* v. 284, p. 1677, 1999.

Weyl, H., *Philosophy of Mathematics and Natural Science,* 1949, Princeton Univ. Press, Princeton.

White, L., *The Evolution of Culture,* 1959, McGraw Hill, New York.

Whitehead, A. N., *Science and the Modern World,* 1925, Macmillan, New York.

Wicken, J. S., *Evolution, Thermodynamics, and Information,* 1987, Oxford Univ. Press, Oxford.

Wiener, N., *Cybernetics: Or Control and Communication in the Animal and the Machine,* 1948, MIT Press, Cambridge, MA.

Wilson, E. O., *The Diversity of Life,* 1992, Harvard Univ. Press, Cambridge, MA.

Woese, C. R., "Universal Phylogenetic Tree in Rooted Form," *Microbiological Reviews,* v. 58, p. 1, 1994.

Wright, R., *The Moral Animal,* 1994, Random House, New York.

# FURTHER READING

Allman, J. M., *Evolving Brains*, 1999, Scientific American Library Books, W.H. Freeman, San Francisco. An excellent, readable and well-illustrated account of how simple organisms that lived nearly a billion years ago evolved the manifold capacities that we humans now enjoy, including big brains having large energy flows.

Anderson, G. M., *Thermodynamics of Natural Systems*, 1996, Wiley & Sons, New York. A pedagogical primer on undergraduate thermodynamics, comparing and contrasting natural systems that are often complex and open with modeled systems that are very much simpler and usually closed.

Atkins, P. W., *The Second Law*, 1984, Scientific American Library Books, New York. A beautifully illustrated book that gives a marvelously instructive introduction to the second law of thermodynamics.

Bertalanffy, L. von, *General System Theory*, 1968, Braziller, New York. A classic interdisciplinary work of the mid-twentieth century, championing a systems theoretic approach to physical, biological, behavioral, and social studies.

Billingham, J. (ed.), *Life in the Universe*, 1981, MIT Press, Cambridge, MA. A fine collection of articles on the cosmic-evolutionary theme and its implications for life elsewhere ("astrobiology") by a highly diverse group of scientists willing to explore an interdisciplinary agenda.

Calvin, M., *Chemical Evolution*, 1969, Oxford Univ. Press, Oxford. Both a top-down (from molecular paleontology) and a bottom-up (from simulated primitive Earth experiments) approach to pre-biological evolution.

Chaisson, E. J., *Universe: An Evolutionary Approach to Astronomy*, 1988, Prentice-Hall, Englewood Cliffs, NJ. An interdisciplinary textbook on cosmic evolution, a nearly verbatim transcript of an undergraduate course taught by the author for the past 25 years, mainly at Harvard University.

——, *Cosmic Dawn: The Origins of Matter and Life*, 1989, W.W. Norton & Co., New York. (First published in 1981 by Atlantic Monthly Press, Boston.) The scenario of cosmic evolution for the general reader, interweaving subject matter from many disciplines to form a unified picture of who we are and whence we came.

——, *The Life Era: Cosmic Selection and Conscious Evolution,* 1989, W.W. Norton & Co., New York. (First published in 1987 by Atlantic Monthly Press, Boston.) A sequel to *Cosmic Dawn,* tracing the history of the concept of change and broaching some of the cosmic-evolutionary ideas developed further in the present book.

Davies, P. C. W., *The Physics of Time Asymmetry,* 1977, Univ. of Calif. Press, Berkeley. An advanced tract, with a full complement of mathematics, that addresses the distinction between time past and time future.

Dawkins, R., *The Blind Watchmaker,* 1986, W.W. Norton, New York. A popular case for genocentric biology, told by an influential thinker who maintains that biological evolution can be understood at the basic level of organic molecules and "selfish" genes.

deDuve, C., *Vital Dust: Life as Cosmic Imperative,* 1995, Basic Books, New York. A cell biologist examines the origins of life, concluding that life and mind are not entirely accidents, rather that life is inevitable and likely widespread given that its basic ingredients seem strewn throughout the Universe.

Depew, D. J. and Weber, B. H., *Darwinism Evolving,* 1995, MIT Press, Cambridge. A scholarly book on the history and philosophy of theories of biological evolution, including modern adjuncts to Darwinism such as self-organization and nonlinear dynamics.

Eigen, M. and Schuster, P., *The Hypercycle—A Principle of Natural Self-organization,* 1979, Springer, Heidelberg. A basic treatise on the theory of the self-organization of matter and of the evolution of biological macromolecules.

Eldredge, N., *Macro-evolutionary Dynamics,* 1989, McGraw-Hill, New York. A first-hand account of the theory of punctuated equilibria, among other recent developments in large-scale thinking regarding refinements to neo-Darwinism.

Emiliani, C., *Planet Earth,* 1992, Cambridge Univ. Press, Cambridge. A thorough, quantitative study of Earth in the cosmos—its geology, evolution, environment, and life.

Feynman, R. P., *QED, The Strange Theory of Matter and Light,* 1985, Princeton Univ. Press, Princeton. A superb, yet unorthodox, account of quantum physics (especially quantum electrodynamics, QED) for a non-technical audience by an acknowledged master.

Goerner, S. J., *Chaos and the Evolving Ecological Universe,* 1994, Gordon & Breach, New York. A new vision of the world as an evolving ecology, this work advocates ecosystems-based thinking as a means to understand scientific and moral order.

Goldstein, M. and Goldstein, I. F., *The Refrigerator and the Universe,* 1993, Harvard Univ. Press, Cambridge, MA. Covers basic, college-level thermodynamics in a readable and entertaining way.

Goodwin, B., *How the Leopard Changed Its Spots,* 1994, Scribner's, New York. A persuasive argument that organisms cannot be reduced to the properties of their genes, but rather must be understood as whole dynamical systems having distinctive properties and intrinsic values characterizing the living state.

Gould, S. J., *Wonderful Life,* 1989, W.W. Norton, New York. An account of the Burgess Shale fossils, emphasizing the role of chance in biodiversity and arguing for a different, non-convergent outcome if the "tape of life" were to be repeated. For a counterpoint, in support of convergent evolution, see Morris 1998, in Works Cited.

Haken, H., *Synergetics* (2nd ed.), 1978, Springer-Verlag, Berlin. An introduction to the mathematics of cooperative phenomena and self-organization in physics, chemistry, and biology.

Harold, F. M., *The Vital Force: A Study in Bioenergetics,* 1986, W.H. Freeman. An especially clear, well-illustrated discussion of the energetics of life forms.

Hawking, S. W., *A Brief History of Time,* 1988, Bantam Books, New York. A surprisingly difficult read for such a best-selling book, but one that touches upon many aspects of the present work: origins, cosmology, time, entropy, thermodynamics.

Ho, M-W., *The Rainbow and the Worm,* 1993, World Scientific Publishing, Singapore. A brief tract on the physics of organisms, generally claiming that a reformulation of thermodynamics is needed to explain life forms—reminiscent of earlier searches for an *élan vital.*

Hofkirchner, W. (ed.), *The Quest for a Unified Theory of Information,* 1999, Gordon & Breach, Amsterdam. Proceedings of the Second International Conference on the Foundations of Information Science, a landmark attempt to use the information sciences to build an integrated understanding of human knowledge.

Jantsch, E. (ed.), *The Evolutionary Vision,* 1981, Westview Press, New York. A disparate group of essays by some mid-twentieth-century pioneers striving to formulate a unifying paradigm of physical, biological, and sociocultural evolution.

Jastrow, R., *Red Giants and White Dwarfs,* 1979, W.W. Norton, New York. A popular account of the origins of stars, planets, and life by one of the modern pioneers of the cosmic-evolutionary scenario. (See also by the same author, *Until the Sun Dies,* 1977, W.W. Norton, New York.)

Kauffman, S., *At Home in the Universe,* 1995, Oxford Univ. Press, Oxford. An unorthodox work of biology, maintaining that life's ordered complexity emerges by means of not only natural selection but also spontaneous self-organization once certain molecules assemble and achieve a supracritical behavior. (A more advanced version of this book is Kauffman 1993, listed in Works Cited.)

Küppers, B-O., *Information and the Origin of Life,* 1990, MIT Press, Cambridge, MA. Argues that natural, Darwinian selection occurs at the pre-biological molecular level, and that biological information arises by the selective self-organization and evolution of biological macromolecules.

Laszlo, E., *Evolution: The Grand Synthesis,* 1987, Shambhala New Science Library, Boston. A sweeping, articulate, and accessible overview of a broad evolutionary synthesis by a noted systems philosopher.

Layzer, D., *Cosmogenesis,* 1990, Oxford Univ. Press, Oxford. An original cosmologist (who prefers a cold start for the Universe) sums up his life's work on the nature of physical laws, the growth of order on all scales, and the origin of cosmogonic systems.

Loewenstein, W. R., *The Touchstone of Life,* 1999, Oxford Univ. Press, Oxford. A discussion of molecular and cellular communication, from the emergence of the first organic molecules to the construction of complex organisms, maintaining that information flow—not energy per se—is the foundation of life.

Lovelock, J., *The Ages of Gaia,* 1988, Oxford Univ. Press, Oxford. Advances the hypothesis that Earth is a non-equilibrium thermodynamic system that optimizes the physical and biological conditions suitable for life by the actions of its own organisms.

Mainzer, K., *Thinking in Complexity,* 1994, Springer-Verlag, Berlin. An intricate discussion of an interdisciplinary methodology explaining the emergence of macroscopic phenomena via non-linear interactions of microscopic elements in complex systems, ranging from natural science to social studies to humanities.

Mason, S. F., *Chemical Evolution,* 1991, Oxford Univ. Press, Oxford. A detailed technical and historical account of both inorganic and organic chemistry, as it pertains to the origin of the elements in the stars and especially to the molecular evolution of minerals and simple organisms on Earth.

Maynard Smith, J., *The Theory of Evolution,* 1993, Cambridge Univ. Press, Cambridge. A comprehensive account of traditional evolutionary thinking in biology today.

Mayr, E., *One Long Argument,* 1991, Harvard Univ. Press, Cambridge, MA. A straightforward explication of Darwin's great idea, and of the subsequent neo-Darwinian synthesis, by one of the twentieth century's foremost evolutionary biologists.

Mercer, E. H., *The Foundations of Biological Theory,* 1981, Wiley, New York. A clear introduction to the physical basis of biological theory; a small text that seeks to unify aspects of natural science.

Miller, J. G., *Living Systems,* 1978, McGraw Hill, New York. A magnum opus by a behavioral scientist who uses general systems theory to analyze the structures and processes for several hierarchical levels of living systems.

Monod, J., *Le Hasard et la Nécessité: Essai sur la Philosophie Naturelle de la Biologie Moderne,* 1970, Editions du Seuil, Paris. (Reprinted as a 1972 English edition, *Chance and Necessity,* Vintage Books, New York.) A thoughtful essay by an eminent biologist for whom life made no sense, endorsing the role of chance and accident in Nature, all the while searching for organic coherence in the gloomiest of existential traditions.

Nicolis, J. S., *Dynamics of Hierarchical Systems,* 1986, Springer-Verlag, Berlin. A primer on the evolution of non-linear systems for the mathematically inclined, stressing both structural and functional aspects of complexity.

Odum, H. T., *Environment, Power, and Society,* 1971, Wiley, New York. A classic, wide-ranging, and readable work on the general systems theory of ecology, by one of its founders.

Prigogine, I., *Introduction to Thermodynamics of Irreversible Processes*, 1961, Wiley, New York. A classical treatise on the physics and mathematics of open, non-equilibrium, thermodynamic systems. (A popularized account, including problems related to philosophy and human sciences, can be found in I. Prigogine and I. Stengers, *Order Out of Chaos*, 1984, Bantam, New York.)

Reeves, H., *Patience dans l'azur: L'évolution cosmique*, 1981, Editions du Seuil, Paris. (Reprinted in a 1984 English edition, *Atoms of Silence*, MIT Press, Cambridge, MA.) Explores the implication of the discovery that the Universe has a history, eloquently using music as a metaphor for Nature's orderliness, and superbly uncovering the many layers of evolution comprising that history.

Reeves, H., deRosnay, J., Coppens, Y. and Simonnet, D., *La Plus Belle Histoire du Monde*, 1996, Editions du Seuil, Paris. (Reprinted in a 1998 English edition, *Origins*, Arcade Publishing, New York.) An insightful three-act play about cosmic evolution for the lay person, compiled from interviews with three leading French scientists—an astrophysicist, an organic chemist, and an anthropologist.

Rue, L., *Everybody's Story*, 1999, State Univ. of New York, N.Y. An insightful and highly accessible exposition of the epic of evolution.

Ruelle, D., *Chance and Chaos*, 1991, Princeton Univ. Press, Princeton. A brief, frank, and well-written introduction to the central role played by the concept of chance in our understanding of the nature of things.

Sagan, C., *Cosmos*, 1980, Random House, New York. A popular, accessible exposition of the big-bang-to-humankind story, and of some of the individuals who contributed to its long journey of discovery.

Salthe, S. N., *Development and Evolution*, 1993, MIT Press, Cambridge, MA. An abstract, idiosyncratic theory of biology stressing the role of predictable development rather than unpredictable evolution as the key to complexity and change in the living world.

Silk, J., *A Short History of the Universe*, 1994, Scientific American Library Books, W.H. Freeman, New York. Articulates well for a general audience the big-bang worldview—the "standard model" of contemporary cosmology.

Simon, H. A., *The Sciences of the Artificial* (3rd ed.), 1996, MIT Press, Cambridge, MA. Considers complexity in the natural world and especially in the "artificial" world caused by human intervention in the natural world—a wide-ranging discussion of mostly cultural evolution by a polymath who taught computer science, economics, and psychology.

Smolin, L., *The Life of the Cosmos*, 1997, Oxford Univ. Press, Oxford. An eloquent exposition by a theoretical physicist who offers a revolutionary view that "cosmological natural selection" among black holes has spawned billions of antecedent universes, each with different physics and perhaps biology.

Spier, F., *The Structure of Big History*, 1996, Amsterdam Univ. Press, Amsterdam. A slim, insightful volume linking social and natural science studies of the past, from big bang until today, stressing cultural advances and written by an interdisciplinarian trained in biochemistry, anthropology, and historical sociology.

Stonier, T., *Information and the Internal Structure of the Universe*, 1990, Springer-Verlag, London. Proposes a new theory that considers information to be a basic property of the Universe, equally alongside matter and energy; partly in accord with that presented in this book, but requiring considerable reformulation of currently accepted physics.

Swimme, B. and Berry, T., *The Universe Story*, 1992, HarperCollins, San Francisco. A well-written exposition of the cosmic-evolutionary narrative for a general audience.

Thompson, D'Arcy, *On Growth and Form* (abr. ed.), 1961, Cambridge Univ. Press, Cambridge. A classic study of biological morphology at the macroscopic level, written with a literary flair and a physico-mathematical orientation, often at the expense of Darwinism.

Tolman, R. C., *Relativity, Thermodynamics, and Cosmology*, 1934, Oxford Univ. Press, Oxford. (Dover ed., 1987.) A classic, if dated, application of thermodynamics to the whole Universe—"an ambitious field of study characterized by danger as well as by interest." Clearly the most mathematically rigorous work cited here.

Ulanowicz, R. E., *Growth and Development*, 1986, Springer-Verlag, Berlin. A theoretical, often formal, treatise suggesting not only that flows of matter and energy generally guide transformations throughout Nature, but especially that an ecosystem's ontological phenomena depend on such flow networks.

Weber, B. H., Depew, D. J., and Smith, J. D. (eds.), *Entropy, Information, and Evolution*, 1988, MIT Press, Cambridge, MA. A thought-provoking collection of diverse articles written by ecologists, cosmologists, biochemists, and philosophers, among others, exploring the interrelations between thermodynamics and evolutionary biology.

Weinberg, S., *The First Three Minutes*, 1977, Basic Books, New York. A classic book that popularized pioneering studies of particle cosmology and the early Universe.

West, B. J., *Lecture Notes on Biomathematics: An Essay on the Importance of Being Nonlinear*, 1985, Springer-Verlag, Berlin. A brief, readable, yet mathematical monograph on the need for non-linear analysis of complex systems in the natural, social, and behavioral sciences.

Wilson, E. O., *Consilience: The Unity of Knowledge*, 1998, Knopf, New York. A leading evolutionary biologist articulates his contention that "all tangible phenomena, from the birth of stars to the workings of social institutions, are based on material processes that are ultimately reducible . . . to the laws of physics."

Wolfram, S., *Cellular Automata and Complexity*, 1994, Addison-Wesley, Reading, Massachusetts. Collected papers of the author, who argues that computer computation will join theory and experiment as methodological equals in the quest to investigate self-organization and chaos in dynamical systems; an alternative to the non-linear, differential-equation approach of dissipative thermodynamics.

# INDEX

# ABOUT THE AUTHOR

Eric J. Chaisson has published more than a hundred scientific articles, most of them in the professional journals. He has also written nine books, including *Astronomy Today* (with S. McMillan), the nation's most widely used college astronomy text-book, and *Cosmic Dawn,* which won the Phi Beta Kappa Prize, the American Institute of Physics Award, and a National Book Award nomination for distinguished science writing. His most recent book, *The Hubble Wars,* also won the Science Writing Award of the American Institute of Physics.

Trained initially in condensed-matter physics, Chaisson received his doctorate in astrophysics from Harvard University, where he spent a decade as a member of Harvard's Faculty of Arts and Sciences. While at the Harvard-Smithsonian Center for Astrophysics, he won fellowships from the National Academy of Sciences and the Sloan Foundation as well as Harvard's Bok and Smith-Weld Prizes. Thereafter, he spent several years as a senior scientist and division head at the Space Telescope Science Institute at Johns Hopkins University, where he was also Adjunct Professor of Physics and Associate Director of the Maryland Space Grant Consortium.

Dr. Chaisson currently holds research professorships in the departments of phys-ics, astronomy, and education at Tufts University, where he directs the Wright Cen-ter for Innovative Science Education; he also teaches at Harvard University and co-directs the Massachusetts Space Grant Consortium at MIT. He lives with his wife and three children in historic Concord, Massachusetts.